Choosing the right NEC contract

Bill Weddell

Published by Thomas Telford Publishing, Thomas Telford Ltd, 1 Heron Quay, London E14 4JD. URL: www.thomastelford.com

Distributors for Thomas Telford books are
USA: ASCE Press, 1801 Alexander Bell Drive, Reston, VA 20191-4400, USA
Japan: Maruzen Co. Ltd, Book Department, 3–10 Nihonbashi 2-chome, Chuo-ku, Tokyo 103
Australia: DA Books and Journals, 648 Whitehorse Road, Mitcham 3132, Victoria

First published 2006

Also available from Thomas Telford Books
NEC Managing reality (complete box set). B. Mitchell and B. Trebes.
ISBN 07277 3397 4
NEC 3 (complete box set). Institution of Civil Engineers NEC Panel.
ISBN 07277 3382 6

A catalogue record for this book is available from the British Library

9 8 7 6 5 4 3 2 1

ISBN: 0 7277 3383 4

Typeset by Academic + Technical, Bristol
Printed and bound in Great Britain by Bell & Bain Limited, Glasgow

Preface

While there have been many changes of construction technology during and since the industrial revolution, the changes in contract strategy, procurement methods and conditions of contract have lagged behind. But the last decades of the twentieth century have seen the beginnings of change – a construction reformation. The appearance of the New Engineering Contract (NEC) and its commendation by Sir Michael Latham in his report of 1994 were undoubtedly an important part of that change. The UK Office of Government Commerce's recommendation of the use of NEC3 by public sector procurers was also significant.

The NEC now has a reasonable track record both in the UK and further afield. This book is the product of the author's experience in advising on and training in the use of the NEC contracts since the beginnings.

Of course there have been problems. Many of these have been caused by:

- Lack of understanding of the NEC contracts: users have often been too ready to import the baggage of the 'traditional' forms of contract.
- Inappropriate drafting of z clauses creating problems of interpretation. The addition of conditions without regard to the standard NEC clauses, usually with a priority clause added, can produce serious problems of interpretation for users.
- Amendments in the form of z clauses that change the allocation of risk to an extent that is tantamount to a change in the nature of the contract. When an 'unreasonable' risk eventuates, the parties quickly revert to traditional combative attitudes.
- Inappropriate choice of contract and of the options available.

The purpose of this book is to help users and potential users avoid some of these problems and provide an overview of the NEC family of contracts. It does not attempt to provide a detailed commentary – that is available in other publications, not least in the guidance notes and flowcharts that have been published to accompany the contract forms. The book also has a positive purpose, namely to help users who may be familiar with the established traditional forms but who are new to the NEC approach to contracts. That is why the content and distinguishing features of each contract are described in each chapter in general terms, and also the choices available in the form of main and secondary options.

Capitals and italics

The conventions of the individual NEC contracts in the use of capital initials for defined terms and italics for terms in the Contract Data have not been followed. Hopefully this will make for easier reading, particularly for those who do not have available the actual contracts. Exceptions are where reference is made to job descriptions or parties with specific roles in a contract. Terms in the Contract Data are sometimes in inverted commas to assist meaning.

Note

The use of him/he has been used throughout rather than the cumbersome him/her, he/she. This in no way reflects the fact that it is a male-only industry.

Acknowledgements

I gratefully acknowledge the contributions of colleagues in the working group, drafting teams and NEC Panel to the debates and discussions (very lively at times) which have taken place and have been crucial in producing the NEC family of contracts. I would also like to acknowledge the helpful discussions with delegates on training courses, both in the UK and abroad, on the NEC contracts. These discussions have usually revolved around the application of the NEC to particular problems and to the difficult issues that are so often part of modern construction. Much of this book is the product of these debates and discussions. The assertions, views and opinions expressed may trespass into controversial areas but nevertheless they are mine alone. They may not represent the opinions of colleagues and others.

Bill Weddell

www.neccontract.com

Contents

Preface

1   The NEC Family   1

2   The Engineering and Construction Contract (ECC)   10

3   The Engineering and Construction Subcontract (ECS)   28

4   The Engineering and Construction Short Contract (ECSC)   36

5   The Engineering and Construction Short Subcontract (ECSS)   44

6   The Professional Services Contract (PSC)   48

7   The Term Service Contract (TSC)   61

8   The Adjudicator's Contract (AC)   75

1 The NEC Family

The beginnings

Although the process of developing a 'New Style Contract', as it was originally called, began in 1985, it was some eight years before the first edition, called the New Engineering Contract, appeared. The concept of a radically different approach to construction contracts was first proposed by the Legal Affairs Committee of the Institution of Civil Engineers. On the basis of the Committee's recommendation, the ICE Council passed a resolution 'to lead a fundamental review of alternative contract strategies for civil engineering design and construction with the objective of identifying the needs for good practice'. With the benefit of hindsight, it can be said that that was a bold step to take at the time and one not devoid of controversy. Standard conditions of contract had been produced and widely used on construction projects for many years. For instance, the ICE had produced the first edition of its standard ICE Conditions of Contract in 1945, immediately after the end of World War II, although many of the contracts used in public works of the nineteenth century reflected the principles and even the terminology used in the first edition. Standard conditions of contract had been produced for use in the building industry much earlier and had been used throughout the twentieth century.

Many construction professionals, with their primary interest in ever-changing technology and its challenges, often regard the legal aspects of their work as of secondary interest, albeit necessary. As a result, they have tended to leave the detailed drafting of conditions of contract to the legal profession. Nevertheless, during the design and construction process, they have been required to become familiar with the powers and duties they have under the contract. It is usually accepted that the best way to understand and learn the conditions of contract is to use them and apply them in practical situations. The result has been that, as practising construction professionals have become familiar with standard conditions of contract and seen them tested and tried with varying degrees of success, they have become reluctant to change. Some have explained this reluctance in terms of natural conservatism and a tendency to stay with the familiar, arguing that engineers and architects are always expected to achieve 100% success. Be that as it may, a radical change to conditions of contract, which have been in use for many years, involves risk and often fear of the unknown. It is with this background that the decision to make what amounted to a step change can be regarded as a courageous one.

The development

The first step was to write a specification for the new-style contract, which included objectives and a statement of what clauses should be included to achieve those objectives. Dr Martin Barnes and Professor John Perry carried out this important seminal task. Drafting of the conditions was not started until the function of the clauses had been clearly identified in the form of 'clause function statements'. During this process, existing standard conditions of contract were reviewed, although the intention of the new-style conditions was to produce a much wider and all-embracing range of contract strategies. Consultation with key figures in the construction industry, particularly employers, followed. Then the process of drafting started. This was done by a small drafting team under the guidance of a working group, which comprised individuals belonging to different professions from across the construction industry. The function of the working group was to decide principles, and the function of the drafting team was to convert those principles into contract clauses.

The drafting began by refining and extending the clause function statements and converting the resulting words into flowcharts. This provided a check on the logic within each clause and, later, on the logic between clauses. It also exposed any overlapping of clause content. Where gaps or inconsistencies were identified, the words of the clause were revised and the flowcharts redrawn and so on, as an iterative process. This process proved to be a powerful discipline, but it resulted in a 'tight', compact and almost fragile set of clauses. One of the consequences of this approach is that any amendment to the conditions is a higher risk exercise than is normally the case; any amendments and additions made to the standard conditions (in the form of so called z clauses – see p. 18) should be carried out with caution, since their effect on other clauses may not be immediately obvious. That is why it is advised, when drafting z clauses, to examine the effect of proposed changes on the flowcharts. The flowcharts, which reflect the final version of the ECC, have been published in a separate document and this practice has been followed for all the contracts in the NEC family. Similarly, guidance notes were drafted in parallel with the text of the conditions and published at the same time.

It is axiomatic that conditions of contract must be legally robust. But the initial structuring of the new-style conditions gave priority to the needs of the technical practitioners in the drawing office and on the site rather than the lawyers in the courtroom. The legal consultations and checks followed the initial process.

Tradition v. Innovation

One of the main issues that the working group and draftsmen had to address in adopting a 'clean-sheet' approach was when and how to depart from traditional practice. On the one hand, the ultimate users would need to be persuaded of the superior merits of any new-style contract if they were to use it. On the other hand, if the new approach were to be too radical, potential users would be unlikely to risk adopting a set of conditions that were untried and untested, however attractive and desirable they may appear to be in theory, with possibly significant consequences. The dilemma constantly posed was: 'To what extent should a lead in the direction of good practice be attempted and to what extent should traditional practice be followed?' Hitherto, development of the traditional standard forms has largely comprised refinement, sometimes as a result of court decisions and sometimes to correct perceived errors or to improve the drafting, but rarely to introduce major changes of principle or policy. Thus, in the new-style contract, in producing a product that would be marketable, the draftsmen had to achieve a balance between tradition and innovation. One example of this dilemma occurred some years later, during the drafting of the Professional Services Contract (PSC). For many years, consultants had often been paid a fee based on the final cost of the works, called an *ad valorem* or percentage fee basis. Although this system of payment was, and still continues to be, widely used, it was rejected as a payment option for a number of reasons. These are fully explained in the guidance notes for the PSC.

Of course innovation for its own sake cannot be justified, but the introduction of change resulting from the need for greater efficiency, value for money and better management of projects in construction, is a serious responsibility for the industry's practitioners.

The three objectives

From the beginning of the project, three objectives were established, namely flexibility, clarity and simplicity, and stimulus to good management. These objectives and principles have been maintained in all subsequent forms of contract comprising the NEC family. They covered areas where most traditional standard forms appeared to be found wanting. They are described in detail in the published guidance notes.

It was considered that flexibility was essential in order to fit the conditions of contract to the precise circumstances of the project or other work in hand. Adaptation has been achieved in traditional contracts by amending the clauses or adding to them. This has often produced problems of interpretation and inappropriate changes in the balance of risk. Flexibility in the NEC contracts has been achieved by producing standard clauses in the form of options. Additionally, fine-tuning of the conditions of contract can be achieved by adding further bespoke clauses, called 'z clauses'. But the provision of so many choices in the options should minimise the need for z clauses. The flexibility available in the NEC makes possible a wide range of payment mechanisms and different allocations of risk.

The reader must judge for himself whether the objective of clarity has been achieved with the aim of producing user-friendly documents. Drafting methods to achieve clarity included the use of short sentences and bullet points, flowcharting, and the avoidance of unnecessary subjective and legal terms. It is significant that the Plain Language Commission awarded an Accreditation Certificate for the first edition in 1999 of the Engineering and Construction Short Contract.

While some standard forms included some aspects of management, it was felt that advances in project management principles and practice, which had taken place in recent times, should be reflected in the conditions of contract. Accordingly, in the NEC contracts, much emphasis is placed on such aspects as the programme, timing and early warning. However, the incorporation of management procedures in conditions of contract itself poses problems. Good management practice demands efficient communication between parties and this, in turn, requires the specification of time periods that are binding on the parties with remedies and sanctions for non-compliance. This raises the vexed question of the status of the stated time periods and whether any are time-barring.

Examination of the time provisions in the various NEC contracts reveals several different categories of which the parties need to be aware. The following categories are taken from the third edition of the Engineering and Construction Contract by way of illustration:

(a) Periods which are clearly time-barring, e.g. clause 61.7 which expressly states that when the defects date is reached compensation events cannot be notified.

(b) Periods which, when not complied with, trigger a compensation event, e.g. clause 13.3 and the related compensation event 60.1(6) for failure to reply to a communication in time.

(c) Periods which, if not complied with, trigger an alternative procedure, e.g. clauses 62.3 and 64.1 for failure to agree the assessment of a compensation event.

(d) Periods which, if not complied with, trigger a remedy either in the same clause or another clause, e.g. clause 51.1 with the remedy in clause 51.2 for failure to certify or pay within the stated time.

(e) Periods which, if not complied with, could invoke the remedy of termination under clause 91.2 reason R11, provided non-compliance can be regarded as a substantial breach of the Contractor's obligations. However, this clause allows the Contractor four weeks to correct the default.

(f) Time-barring periods, e.g. clauses 61.3, 61.4 and 62.6, some of which incorporate a 'period of grace' to allow timely compliance.

Terminology

One important matter that the drafting team had to address was the terms to be used in the new-style contract. The danger of using traditional terms was that they might be assumed to have the same meanings and be associated with the same procedures as those in traditional contracts. The arguments in favour of using traditional terms include the fact that they are well understood and that in many instances they are supported by decisions of the courts. However, it has been stated that in relation to the ICE Conditions of Contract remarkably few court cases have revolved around the meaning of the terms. Different meanings suggest the need for using different terms. While superficial reading of the contract may suggest little difference in the meaning of some terms, close examination will demonstrate differences.

The first edition of the New Engineering Contract

A consultative edition of the NEC was published in 1991, some five years after the ICE had decided to proceed with the new venture. The reason for issuing a contract for consultation was clearly to obtain comments from as wide a section of the industry as possible. Some Employers evidently felt sufficiently confident to use the consultative edition in real situations. Feedback from these contracts, a number of which were in countries other than the UK, was extremely valuable.

Comments were considered and debated by the working group and drafting team, which then produced the first edition for publication in 1993.

A consultative edition of a standard subcontract form for use with the NEC was published in 1991, at the same time as the publication of the consultative edition of the NEC. This was followed by the first edition of the subcontract form in 1993. It was described as being back-to-back with the NEC. A glance at the subcontract form shows that the wording of the two forms is identical except in places where the differences in nature of the two contracts demand different wording and slightly different provisions. This contrasts with the ICE Conditions of Contract where a subcontract form was produced by a different organisation – the contractors' organisation the Federation of Civil Engineering Contractors (FCEC), later the Civil Engineering Contractors' Association (CECA).

Expansion of the NEC family

The need for a compatible subcontract form was evident during the early days of the development of the new-style contract, and it soon became obvious that other standard forms were needed. The NEC had introduced a Project Manager and a Supervisor to fulfil some of the similar roles of the Engineer of the ICE Conditions of Contract. The logical next step was to produce a standard contract for the appointment of such professional people, using the same principles as were used in producing the NEC – in other words, a standard form for the appointment of Consultants. Many standard forms for the appointment of Consultants have been produced unilaterally by either Employers' or Consultants' organisations. It was therefore sensible that all the principles used in producing the NEC should also be used in drafting a standard contract for the appointment of Consultants. In 1993, soon after the publication of the first edition of the NEC, a consultative edition of a Professional Services Contract (PSC) was produced and this was followed in 1994 by a first edition.

The Latham Report

The publication in July 1994 of Sir Michael Latham's report on the construction industry was a significant event. This report, 'Constructing the Team', had been commissioned jointly by both government and industry. It was probably the most significant report on the construction industry since the Banwell Report of 1964. Among the large number of wide-ranging recommendations, the Latham Report listed some 13 matters that should be included in any effective form of construction contract under modern conditions, in order to promote best practice. Latham described the approach of the New Engineering Contract as being 'extremely attractive' and commended it as containing most of the assumptions of best practice. This represented an important endorsement of the NEC approach to contracts. The report did, however, suggest a number of changes to the NEC. These were considered by the NEC Working Group and incorporated in the second edition together with other changes resulting from

Table 1.1. Published NEC contracts

Title	Abbreviation	Consultative edition	First edition	Second edition	Third edition
New Engineering Contract/ Engineering and Construction Contract	NEC/ECC	Jan. 1991	March 1993	Nov. 1995	June 2005 (NEC3)
NEC Subcontract/ Engineering and Construction Subcontract	NEC Subcontract/ECS	Jan. 1991	March 1993	Nov. 1995	June 2005 (NEC3)
Professional Services Contract	PSC	June 1993	Sep. 1994	June 1998	June 2005 (NEC3)
Engineering and Construction Short Contract	ECSC	Feb. 1997	July 1999	June 2005 (NEC3)	
Engineering and Construction Short Subcontract	ECSS		July 2001	June 2005 (NEC3)	
Adjudicator's Contract	AC	June 1993	1994	June 1998	June 2005 (NEC3)
Term Service Contract	TSC	May 2002	June 2005 (NEC3)		
*Framework Contract			June 2005 (NEC3)		

*The Framework Contract is not discussed in this book. All contracts were either revised or published for the first time in June 2005 under the heading NEC3

feedback from, and further consultation with, users. One of the changes was to the title of the contract in order to reflect more accurately its application throughout all sectors of construction – building, mechanical, electrical and process engineering, as well as civil engineering. The second edition was published in 1995 under the new title of the 'Engineering and Construction Contract'. The acronym NEC has been retained to describe the family of standard contracts that have been developed based on the same principles. In fact, the Latham Report also recommended production of 'a complete family of documents', which should include 'a full matrix of consultants' and adjudicators' terms of appointment'. As described above, this process had already commenced by the publication of the subcontract version of the NEC, the Professional Services Contract and the Adjudicator's Contract. Since then, the family of NEC contracts has continued to expand in response to industry needs and the Latham recommendations. Details of the NEC standard conditions of contract, available at the time of writing, are given in Table 1.

Choosing the right NEC contract

The purpose of this book is to inform those Employers who believe that the NEC approach to carrying out a project is the most appropriate one, and to assist them in deciding how to proceed. Each of the NEC contracts, other than the Framework Contract, is described in subsequent chapters. Each contract is described in general terms, emphasising those features that influence the decisions and choices to be made – usually by the Employer, but in the case of subcontracts, the Contractor. Also included are a brief history of the origins of each contract, an overview of its contents, what it is intended to be used for and how to use it. The chapters do not attempt to include a detailed description or commentary on every clause, since this information is available in the published guidance notes and other publications. Rather, the more important provisions of each contract are described, including those features which distinguish the NEC contracts from the more traditional contracts. It is important not only to select the NEC contract most suited to the Employer's objectives but also to select the most appropriate options available within the contract. Hence, a description of the various main and secondary options is included. A thorough understanding of these is fundamental, so that appropriate decisions can be made to suit the circumstances for any particular project. It is possible that this book may be used as a reference manual and thus an attempt has been made to make each chapter largely self-contained. As a consequence, there is a certain amount of repetition between chapters. Some repetition is avoided by reference to other chapters.

Choosing the right NEC contracts for construction management

Construction management is to be distinguished from management contracting, in that it is the Employer, or a construction manager acting on behalf of the Employer, who manages the employment of different contractors on a project. The Employer is a party to each contract, and the construction manager's function is to manage the physical interfaces between each contract. He also manages the all-important time interfaces, usually by means of a master programme, to ensure co-ordination and smooth running of all activities. The flexibility available (described in subsequent chapters) in assessing the effects of change (compensation events) in each contract is a major factor in facilitating the effective management of time interfaces. Also, the introduction of key dates in the NEC3 contracts provides a further mechanism by which different contracts can be successfully managed.

Choosing the right NEC subcontracts for management contracting

A management contract is one in which a Contractor is appointed – the Management Contractor – whose main function is to manage the works packages that are let as subcontracts. Thus the managing of the interfaces between the subcontracts is entirely the responsibility of the Contractor. For a management contract, the main NEC contract is placed under option F of the Engineering and Construction Contract (see Chapter 2). The selection of the appropriate form of subcontract will depend on a number of factors. Depending on the nature of the work comprising each subcontract works package, the Management Contractor may select the Engineering and Construction Subcontract (ECS), the Engineering and Construction Short Subcontract (ECSS) or, where design work is required, the Professional Services Contract (PSC). The choice of subcontract may also be subject to any requirements specified by the Employer under the main contract. The Employer would also normally specify procurement requirements. The choice of main and secondary options under the ECS and PSC will depend on the considerations discussed in the chapters on those contracts.

All construction contracts involve risks that must be identified, minimised and then allocated

In recent years, risk has been increasingly recognised as a major factor in construction. In the past, there is little doubt that inadequate consideration of risk has been the cause of many of the industry's problems. Risk has been the subject of much study and it is a vital factor in choosing not only the right contract but also the right NEC contract, which is the subject of this book.

In any construction project there are many risks. They vary widely in nature, in the probability that they will occur, and in the consequences when they do occur. For example, one of the major risks in tunnel construction is the nature of the ground; changes in ground conditions may have a profound impact on the cost of construction. By way of contrast, the risk of unexpected ground conditions when building a multi-storey hotel is much smaller since the cost of ground investigation on a relatively small plan area is not great. Inflation is a further example of a risk that may vary widely. In some countries with a record of high inflation, the risk of increases in prices may be high even over short periods of time. In other countries, with stable economies, the risk may be much smaller but even then, where a contract period may be, say, ten years (now increasingly common for long-term maintenance contracts) the risk of inflation may be high.

In the past there has been a tendency on the part of many major Employers, who decide the form of contract and the allocation of risk, to require the

Contractor to carry as many risks as possible. While superficially this has many attractions for the Employer, it places tendering contractors in a difficult position in pricing their bids. This applies particularly where a number of contractors tender in competition and the selected contractor is usually the one that has submitted the lowest tender. The result has often been that the successful contractor is the one with the most efficient claims department rather than the one that is the most proficient in technological and management terms. Contractors have often been forced to become 'claims conscious' in a competitive market by reason of the prevailing system and the standard conditions of contract. In other words, this policy has usually led to the rewards going to the most litigious; it accounts for many of the disputes and frequent confrontation that have occurred, resulting in increased construction costs, delay and legal costs.

In the UK, for many years, much emphasis has been placed on freedom of the parties to contract. It is only rarely that Parliament has intervened with this freedom, usually to provide some protection where the bargaining powers of the parties are widely different. Thus the UK courts will normally concentrate on the literal meaning of the words of a contract rather then on the reasonableness or otherwise of its terms. However, to avoid the situation described above, it is sensible to allocate as 'fairly' as possible the risks in the project. This requires a careful identification of the risks, examination of how the risks can be eliminated or at least reduced, and then allocation of the remaining risks to the party best able to manage them. It is significant that the NEC3 family of documents has introduced a 'risk register' with the object of furthering the process of managing risk.

There are many examples of 'unfair' or 'unreasonable' risks which Employers have sought to place on the Contractor, sometimes with unfortunate consequences. These include delays caused by the Employer, as, for instance, where information is provided late; where the Employer has designed the works but nevertheless made the Contractor responsible for checking the design and making him liable for design errors; allocating to the Contractor liability for the consequences of discovering existing services in the ground; allocating to the Contractor liability for the performance of third parties such as service authorities involved in the contract; risks outside the control of both parties. Of course, one risk which the Employer cannot transfer to the Contractor is the ultimate one of liquidation or bankruptcy of the Contractor. That is not in the interests of any party.

While it is not in the interests of the contracting parties or of the project itself to allocate too much risk to the Contractor, conversely it is not desirable to place all risk on the Employer. That is likely to have the effect of removing from the Contractor the incentive to work efficiently. Thus, for any project, there is likely to be an optimum allocation of risk.

All contract forms in the NEC family provide options

In preparing a particular contract, decisions must be made to achieve the optimum allocation of risk. All the standard NEC contracts, other than the short forms and the Adjudicator's Contract, contain main options and secondary options. One of the most important decisions which the Employer must make, is to decide which main option to use. This determines the financial risk to be taken by the Contractor and also how he is to be paid. For example, in the ECC there is a gradual shift of risk from option A, where the Contractor carries most of the financial risk, to the cost reimbursable option E, and also F (the management contract) where the Employer carries most of the financial risk.

Also included are a number of secondary options. Selection of these is largely determined by considerations of the optimum allocation of risk described above. The choice of which dispute resolution option W1 or W2 (which is a feature of the NEC3 contracts) to include depends on whether the contract is subject to the UK's Housing Grants, Construction and Regeneration Act 1996; thus this choice should be straightforward.

Further flexibility is provided in the z clauses, which permit the Employer to add his own bespoke clauses to the standard clauses. The published guidance notes provide some limited advice on z clauses but the main principle to follow is that they should not upset the main balance of risk inherent in the selected main and secondary option clauses.

Exercises

(1) What problems currently exist in the construction industry? Which of these do you consider could be solved by appropriate conditions of contract and how?

(2) An Employer requires to construct a pedestrian tunnel under a tidal river. He wishes to place all risks on the Contractor and has expressed the view that he is prepared to pay a higher price so that he can achieve this. He has appointed you as professional adviser. What advice would you give him and what reasons for the advice would you give?

(3) 'Conditions of contract should be kept in the drawer and only consulted when a problem arises.' Discuss.

(4) A major new hospital is proposed on the site of an old gas works, which was disbanded 20 years previously. The site is in a residential area. List the major risks on such a project and describe how they might be eliminated or minimised, making any necessary assumptions. How would you allocate the remaining risks between the Employer and the Contractor?

2 The Engineering and Construction Contract (ECC)

How the ECC began

The consultative and first editions A consultative edition of the New Engineering Contract was published in January 1991. This was accompanied by extensive guidance notes and a full set of flowcharts used during the drafting process. The purpose of publishing a consultative version was to give an opportunity to those who were seeking a new approach to construction contracts, to try it in an actual situation and to test it to see whether the principles on which it had been drafted were valid in the real world of contracting. Comments were also invited from the industry generally. During the consultation period many comments were received and considered by the NEC Working Group and drafting team. The result was the publication of the first edition in March 1993.

The second edition The second edition was published in November 1995. The main factor that led to a revision so soon after the first edition was the publication of the Latham Report, 'Constructing the Team', in July 1994 as described in the previous chapter. The second edition was named the Engineering and Construction Contract. This change was in response to the recommendation of the Latham Report to change the title to one that represented its application across the whole of construction. Other changes were made as a result of further use and experience of the contract as well as the Latham recommendations.

The third edition The ten years after the publication of the second edition saw increasing use of the ECC in all sectors of the industry and in a number of other countries. This increasing use probably arose from a number of factors including:

- major changes in methods of procurement;
- the wide use of partnering in both public and private sectors;
- the increase in design and build contracts, often leading to maintenance of the asset by the contractor;
- privately financed projects;
- perceived benefits in collaboration of the contracting parties rather than confrontation, with a consequent reduction in disputes;
- recognition of the benefits of a 'fairer' allocation of risk; and
- disillusion with over-runs of cost and delays in completion using traditional forms of contract.

The NEC Panel reviewed the performance of the contract over this period and investigated how to incorporate developments in best practice that had occurred during the intervening period. This resulted in the publication of the third edition under the general title NEC3 in June 2005. It is significant that this edition received the official endorsement of the UK's Office of Government Commerce (OGC) in that it complies fully with the *Achieving Excellence in Construction* (AEC) principles. Accordingly, the OGC recommends the use of NEC3 by public sector construction procurers on construction projects.

The essential and distinguishing features of the ECC

The published documents One of the problems with traditional standard forms is that they rarely meet the precise requirements and circumstances of a particular project. The result is that they are often used in situations that are inappropriate. This may result in high risk being placed on the contractor, leading to disputes when the risks eventuate. Alternatively, the standard conditions may be amended to deal with the particular circumstances. This commonly produces inconsistencies, uncertainty of meaning and confusion, frequently ending in disputes and much employment for lawyers.

The published ECC document is not a contract but rather a number of clauses and statements from which a contract to suit particular circumstances may be prepared. Hence, the preparation of tender and contract documents requires considerable thought and effort; choices have to be made in order to produce an optimum and appropriate set of conditions of contract. These choices, usually made by the employer, are crucial to the successful outcome of the contract. A contract under the ECC will thus comprise:

(a) Core clauses. These must be included in every contract.

(b) One main option selected from the six main options A to F. Selection of the main option determines the allocation of risk between the parties and how the Contractor is paid. The numbering of the clauses follows on from the core clauses and under the same side headings as in the core clauses. Separate documents have been published containing the conditions for contracts using each of the main options.

(c) A dispute resolution option W1 or W2, according to whether the UK's Housing Grants, Construction and Regeneration Act 1996 applies to the contract.

(d) Secondary options selected from the numbered options X1 to X20 (though some numbers have not been used in the ECC), and the two options numbered Y(UK)2 and Y(UK)3. This selection determines the allocation of further risks.

(e) Additional conditions of contract indicated by the letter 'z'. These are normally decided by the employer or by agreement between the parties. Because of the options available, the need for additional conditions should be minimal.

(f) Contract Data. This is in two parts, part 1 being prepared by the employer and part 2 prepared in part by the employer for completion by tenderers or the selected contractor. The Contract Data contains information specific to the contract. Some information in the Contract Data takes the form of reference to other documents, which are thereby incorporated into the contract.

One of the most important decisions the employer has to make is to decide which main option to use

There are six main options, as below.

The Contractor carries the greatest financial risk under options A and B, and the least risk under options E and F. Selection is determined by the nature of the work to be carried out, and how the risks are to be dealt with.

The lump sum option. Option A: Priced contract with activity schedule

This is basically a lump sum contract in which the Contractor offers to provide the works for a sum of money. He takes the risk of being able to produce the works described in the contract for the lump sum and hopefully make a profit. This, of course, is subject to the terms of the contract, which provide for certain risks to be carried by the Employer. The activity schedule is a list of activities to be carried out by the Contractor in providing the works. The Contractor normally writes this since he is the one who knows what activities will be carried out. However, the Employer may specify the form of the activity schedule. Each activity is priced by the Contractor as a lump sum, which is the amount paid to the Contractor when he has completed the activity. Because of this payment condition, contractors should ensure that the number and description of items in the schedule are such as will maintain a reasonable cash flow. The total of the lump sum prices is the Contractor's price for completing the whole of the works. In pricing an activity, the Contractor takes responsibility for estimating quantities and resources, and assessing and pricing the risks that are his.

© copyright nec 2006   11

The remeasurement option. Option B: Priced contract with bill of quantities

A bill of quantities is a list of work items and quantities. It is prepared by the Employer in accordance with detailed rules contained in a 'method of measurement'. This is usually a standard published document that states the items to be included, how the work described in the item is to be measured, how the quantities are to be calculated and sometimes what is to be allowed for in the rate for each item (item coverage). The bill may include lump sum items as well as quantity-related items. It may also include time-related items. The Contractor prices each item. The total amount for each quantity related item is calculated by multiplying the estimated quantity by the rate. If it is found that the quantity is not correct when the work is done, it is corrected and payment to the Contractor is adjusted to reflect the actual work carried out. The need for this correction of quantities may be due to errors in the original calculation or because actual quantities of work done differs from the estimated quantities. Hence, under this option, unlike option A, the Employer takes the risk of the correctness of the quantities. The process of correction is sometimes called admeasurement or remeasurement. Payment of the Contractor is on the basis of the quantity of work done – he does not have to wait until an item of work is completed as is the case under option A.

Bills of quantities have been used widely in construction work in the UK for many years. Option B is clearly not appropriate for design and build contracts since it is the Contractor who designs and prepares the detailed drawings, not the Employer.

Option B should be used where the risk of change in quantities is relatively high. This may be the case where earthworks comprise a significant part of the contract and the material excavated may fall into different categories. There may be different items in the bill for each category of excavated material and filling material.

The target option with the target established as a lump sum. Option C: Target contract with activity schedule

Under this option, the contractor tenders or negotiates a target price (defined as the Prices) using an activity schedule. He also tenders a Fee in terms of fee percentages (one fee percentage is for subcontract work and one for the Contractor's own direct work). The Fee broadly covers the Contractor's head office overheads and profit, disallowed costs and any other cost not included in Defined Cost. During the course of the contract the Contractor is paid his actual cost (called Defined Cost in the contract) plus the Fee. An important change in NEC3 is that payment is made on the basis of a forecast of payments made by the Contractor before the following assessment date, rather than on payments that have already been made. At the end of the contract, the final Defined Cost plus the Fee is compared with the target. If this shows a saving, the difference is shared between the Employer and Contractor in proportions that are defined in the contract. If it shows that the target price has been exceeded, the Contractor must pay back a proportion of the excess. In fact the 'Contractor's share' is calculated on two occasions – at Completion, when the calculation is based on estimated final figures (for target price and Defined Cost), and at the final account stage, when the calculation of the Contractor's share is refined using final figures. During the course of the contract, the target price is adjusted to cater for compensation events – these are events that are described in the contract and are at the Employer's risk.

There is one provision in this option (and also in option D) that gives the Contractor an incentive to suggest possible changes to the Employer's design that save cost. In effect, this is a value engineering clause. It provides that if such a proposal by the Contractor is accepted by the Project Manager, the target (the Prices) is not changed. This will result in an increase in the Contractor's share and a saving in cost for the Employer – a 'win–win' situation.

Option C is basically a cost reimbursable contract, which incorporates an incentive for the Contractor to minimise costs. Savings and over-runs are shared between the parties. Since its first introduction in the first edition, the target option has increased in popularity. This is probably because, in many projects, risks are relatively high, such that a priced contract (under Options A or B) could produce many changes with an increased likelihood of disputes arising from pricing the changes. The sharing of risk, as in a target contract, is likely to reduce the occurrence of disputes. The extra administrative effort required in a target contract should not be under-estimated since it requires constant calculation of Contractor's costs as well as assessment of compensation events for the purposes of adjusting the target.

The target option with the target established as a remeasurable sum. Option D: Target contract with bill of quantities

This is similar to option C except that the target price is established by means of a bill of quantities instead of an activity schedule. During the course of the contract, the target price is adjusted to allow for changes of quantities as well as for compensation events. Thus, the Employer carries a rather greater risk than is the case with option C. This option should be used in preference to option C, where the likelihood of change of quantities in the bill is relatively high. The only use of the bill of quantities in an Option D contract is for the purpose of adjusting the target, not for calculating the periodic payments to be made to the Contractor.

The option where the Contractor is paid his costs. Option E: Cost reimbursable contract

Under this option the Contractor takes a very small risk since he is paid his actual cost (i.e. Defined Cost, which includes a deduction for 'Disallowed Cost') plus the Fee with only a small number of constraints to protect the Employer from the effects of incompetence and inefficient working on the part of the Contractor. It is used when the work to be carried out cannot be easily defined at the outset and when the risks in doing the work are great. It may also be used for urgent and emergency work where preparation of detailed contract documents and proposals for design are precluded. The option is also suitable for experimental and research projects, where decisions on what is required may have to be made on a day-to-day basis.

The option where the Contractor is a Management Contractor who manages the work to be done. This work is broken down into work packages, which are carried out by Subcontractors. Option F: Management contract

This option is suitable for management contracts in which all or most of the construction work is done by subcontractors, and the Contractor manages the procurement and carrying out of the subcontract works. Payment of the Contractor consists of payments to subcontractors plus the Fee which is effectively a management fee. Thus, the Employer carries most of the risk in a management contract. The risks remaining with the Contractor are broadly the level of his fee percentages and elements of 'Disallowed Cost' as defined in the contract.

The dispute resolution options that determine the procedures for dealing with disputes

The dispute resolution options W1 and W2

Although the parties to the contract are required to act in a spirit of mutual trust and co-operation, and are thereby encouraged to settle their differences without recourse to a third party, such are the complexities and pressures of modern construction that some disputes are almost inevitable. The method of resolving disputes in the first instance, in most traditional contracts, for many years has been arbitration. The consultative edition and subsequent editions of the NEC introduced adjudication as the primary contractual method of resolving disputes. Among the reasons for this change was the fact that arbitration had become both time-consuming and expensive. The result has been that justice has not always been done and the claimant has sometimes been reluctant to start the process because of the high legal and other costs involved. Further, Latham recommended that adjudication should be the normal method of dispute resolution and that this requirement should be underpinned by legislation. In the UK the right to refer any dispute to adjudication was enshrined in the Housing Grants, Construction and Regeneration Act 1996 (the Act). However the Act applies only to construction contracts as defined in the Act and only to contracts in the UK. Thus NEC3 has introduced two dispute resolution options:

- Option W1 for contracts not subject to the Act.
- Option W2 for contracts that are subject to the Act.

To decide which option to use, it is necessary to check whether a contract is a 'construction contract' as defined in the Act. This definition is stated in terms of 'construction operations'. Most of the clauses in the two options are identical but some in W2 are a specific requirement of the Act. These have not been included in W1 in order to manage contracts not subject to the Act more effectively.

The main provisions of option W1 are:

- The adjudicator is named in the Contract Data and the parties enter into a contract with the named person under the NEC Adjudicator's Contract.
- The adjudicator acts impartially.
- The adjudicator is not liable to the parties unless he acts in bad faith.
- There are four categories of dispute, each with stated constraints on the time of referral.
- Provision for joining a subcontract dispute on a certain matter with a dispute under the main contract involving the same matter, and have it decided by the main contract adjudicator.
- The adjudicator has wide inquisitorial powers of investigation, etc.
- The adjudicator decides the dispute within four weeks and the decision is enforceable in the courts.
- A dissatisfied party may refer the disputed matter to the tribunal (named in the Contract Data).

Provisions that are specific to option W2 include:

- Disputes may be referred to an adjudicator at any time.
- A procedure where a party fails to comply with an adjudicator's instruction.

The Adjudicator's decision is not final, in that a party dissatisfied with the Adjudicator's decision can refer the dispute to the tribunal, which is specified in the Contract Data as (normally) arbitration or litigation in the courts. The Adjudicator is named in the Contract Data, is appointed jointly by the parties, and is brought into action only when a dispute is referred to him.

Various further risks need to be considered to decide how they can be reduced and which party should carry them. This is the function of the secondary options

Which party should carry the risk of price increases? Option X1: Price adjustment for inflation

In countries where the rate of inflation is high, the Employer must make the decision on which party is to carry the risk of inflation. If the Contractor takes this risk, he must assess the effect of price increases occurring during the period of the contract and allow for them in pricing the tender. This is a high-risk exercise since inflation is usually a matter of economics and politics, and has nothing to do with the efficiency and technical competence of the Contractor. Thus, if this option is not included, the Employer may be paying an excessive amount for the work. Alternatively, the Contractor may lose excessive amounts with a high risk of liquidation, resulting in problems for the Employer. Inclusion of this option is therefore advisable where inflation is likely to be high or variable, and in cases of relatively long contract periods.

Inclusion of this option effectively transfers the risk, or at least most of the risk of inflationary increases in prices to the Employer. The method of calculating adjustments to the prices is by means of a formula, which makes use of indices for labour and materials. These indices are published nationally in many countries. The particular indices selected should be those that reflect the nature and content of the works. In countries where no such indices are published, it would not be possible to use this option; an alternative provision would be required to ensure that the Contractor is reimbursed for changes in prices.

In the case of options A and B the adjustment is made to the amount otherwise due to the Contractor as calculated from the activity schedule or bill of quantities. For options C and D adjustment is made only to the target (the Prices) as the Contractor is paid his actual costs (Defined Cost plus Fee) periodically throughout the contract and this, therefore, includes any price increases that may have occurred. In options E and F the Employer automatically carries the risk of inflation.

Which party should carry the risk when there is a change in the law? Option X2: Changes in the law

During the period of a contract, a change in the law may affect a contractor's costs. In some countries such a change, for example a change in import duties or customs payments, can have a major effect on a contractor's costs. Inclusion of this option transfers the risk of the effects of changes in the law from the Contractor to the Employer. It applies whether the changes in the law have the effect of increasing or reducing the Contractor's costs.

If the Contractor is to be paid in more than one currency, how is this done and which party takes the risk of changes in the exchange rates? Option X3: Multiple currencies

This option is intended for use on international contracts where the Contractor may require to be paid in more than one currency. A 'currency of this contract' is established for each contract and stated in the Contract Data part 1. This option is designed for use only with options A and B. The Contractor prices his offer in the 'currency of this contract' and the items of work that are to be paid for in a different currency are listed in the Contract Data. Conversion to other currencies is done by using the 'exchange rates' stated in the Contract Data. These are fixed at a definite date (usually some weeks before the Contract Date), which is also stated in the Contract Data.

In options C to F the Contractor is paid on the basis of his actual costs. Thus, payment is made in whatever currency the Contractor incurs the cost. This is made clear in the clauses for these options under the heading of multiple currencies.

What security does the Employer require in the event of the Contractor's failure to carry out his obligations? Option X4: Parent company guarantee

Where the Contractor is owned by a parent group of companies, the inclusion of this option gives the Employer greater security of the Contractor's performance in the form of a guarantee by the parent company. The form of the guarantee is stated in the contract. Failure by the Contractor to provide the guarantee within stated time periods entitles the Employer to terminate the contract.

Does the Employer require the works to be completed in sections? Option X5: Sectional Completion

The Employer can include this option when he requires a section of the works to be completed before the whole of the works. The sections are described in the Contract Data together with their respective completion dates. The sections do not make up the whole of the works but provide only for stated completion dates before completion of the whole of the works. Sectional completion should be used where the Employer requires the Contractor to complete a defined physical part of the works so that he can take it over. It should not be confused with the system of Key Dates under which the Contractor is required to carry out the works to reach a stated condition within a programmed timescale. Key Dates are most commonly used on projects employing a number of contractors each of which is dependent on completion of work to a certain stage by others.

Option X5 may be used with option X6 where the Employer wishes to include an incentive to complete a section of the works as soon as possible. Similarly, the option may be used with option X7 to include for delay damages payable by the Contractor for failure to complete a section by its completion date.

Does the Employer want the works to be completed as soon as possible? Option X6: Bonus for early Completion

In some contracts, the Employer may wish to have the works (or a section of the works) completed at the earliest possible date. This option provides a strong incentive to the Contractor to complete early. For each day he saves in construction time, the Contractor is paid a bonus as stated in the Contract Data.

What happens if the Contractor fails to complete the works on time? Option X7: Delay damages

Failure by the Contractor to complete the works by the Completion Date is a breach of contract by the Contractor for which the Employer is entitled to some recompense. This option provides for the quantification of this recompense in the form of a sum of money for each day on which the Contractor is late. In many standard forms of contract under English law, delay damages are described as liquidated damages. The amount of delay damages is stated in the Contract Data and must be no greater than a genuine estimate of the financial damage suffered by the Employer as a result of late completion. If this option is not included, delay damages are unliquidated or 'at large' and recovery of the Employer's losses can be much more difficult to evaluate and recover.

This option can be combined with Option X5 (Sectional Completion) to provide the Employer with a remedy in the event that particular sections of the work are not completed by their respective completion dates. In this case an estimate of delay damages should be stated in the Contract Data against the completion date for each section.

Where the Employer takes over and uses a part of the works before Completion, the delay damages are reduced proportionately to reflect the benefit which he has gained.

The parties may wish to cap the total delay damages that are payable. This can be achieved by inserting an appropriate statement in the Contract Data.

How can the Contractor be encouraged to work with other parties as a team, on the project, to produce a successful outcome? Option X12: Partnering

One of the main intentions of the NEC provisions is to encourage collaborative relationships between the contracting parties rather than creating an environment of confrontation as is the case with many traditional forms of contract. Partnering has been introduced in recent years to encourage collaboration in a similar way but between all parties who may be involved in a project or series of projects. When partnering was first introduced, the arrangement between the various parties was non-contractual. However, there arose a demand for such arrangements to be enshrined in a legally binding contract or contracts. To many, this seemed to be an attempt to combine obligations that were contradictory, but this view may have resulted from the confrontational nature of traditional contracts. However, in response to the demand, the Construction Industry Council (CIC) published a 'Guide to Project Team Partnering'. This was not a partnering contract but merely provided guidance on matters that were considered to be essential to a partnering contract.

Since the NEC family of contracts is based on a policy of co-operation rather than confrontation, all that was necessary to create a true partnership was to devise a mechanism for creating the necessary relationships between the various parties already operating under one of the NEC contracts. This is the function and effect of option X12 that is added to each individual NEC contract to create these relationships. The partners are listed on a Schedule of Partners, which is constantly updated and revised as new partners join the project and others leave having completed the work in their contracts. A core group of partners acts as an executive body on behalf of all partners. The members of the core group are chosen by the partners and the group is normally led by the client representative, the client being the Employer in the individual NEC contracts.

Details of how the partners are to work together are stated in the document described as Partnering Information. The partners are required to provide and exchange information that is needed for their purposes and to use common information systems. In addition, the partners are required to provide advice to each other when requested by the core group. The core group has the authority to change the Partnering Information but such a change constitutes a compensation event under the individual partner's contract, that is, it may result in extra or reduced payment and delay to the completion date under the individual contract.

An important aspect of partnering is the co-ordination of the activities of the partners. This is achieved by means of a master programme (called a timetable). Any change to the timetable will usually affect a partner's programme in his individual contract, and thus such an imposed change constitutes a compensation event.

Incentives to improve the performance of partners both individually and collectively are catered for in the form of Key Performance Indicators (KPIs).

How can the Employer obtain some financial security when the Contractor fails to perform? Option X13: Performance bond

In most construction contracts, the employer requires some form of security as a protection against the contractor's failure to perform his obligations for whatever reason. Under the ECC this takes the form of a performance bond provided by a bank or insurance company. In effect this is a method of transferring the risk of a Contractor's non-performance to a third party. The option does not specify the form which the bond is to take – this is to be stated in the Works Information. Thus, the conditions that must be fulfilled before the Employer may call the bond, may be stated in the Works Information, or he may elect to require the less common unconditional or 'on-demand' bond. Failure to provide the bond is regarded as a serious breach, which may give rise to the ultimate sanction of termination.

How can the Employer help to finance the contract during the early months? Option X14: Advanced payment to the Contractor

In cash flow terms, the most difficult period for a contractor is at the start of a contract. Because of the need to mobilise resources soon after the award of a contract, often involving the mobilisation and import of personnel and major items of equipment, a contractor may expend considerable sums of money before he receives any payment from the employer. To alleviate this financial burden, provision is sometimes made for a payment to the Contractor in advance of work being carried out. To this extent, part of the financing of the contract is transferred from the Contractor to the Employer and this should result in lower contract prices. This option provides for such an advanced payment. The payment is made soon after the contract date and is repaid in stages progressively throughout a stated period. There is provision in the option for a bond to be provided by the Contractor to protect the Employer against non-repayment of the moneys that have been advanced. This bond is not compulsory but would normally be required by the Employer. As in the case of the performance bond, failure to provide the bond within the stated time gives grounds for termination and, conversely, failure by the Employer to make the advance payment by the stated time constitutes a compensation event.

Should there be some restriction of the Contractor's legal liability for design work which he carries out? Option X15: Limitation of the Contractor's liability for his design to reasonable skill and care

This option relates only to design work done by the Contractor and thus would not be used where the Employer has designed the whole of the works. Basically, the Contractor's obligation is to design in accordance with the Works Information. The latter document should state what the Contractor must design and may include how he is to design, that is, a statement of any design and performance criteria with which the design must comply. Without this option, the Contractor's obligation is an absolute one in that the design must comply with the Works Information. The introduction of this option lowers the standard of care required of the Contractor for his design work, from fitness for purpose (as in core), to reasonable skill and care, which is the normal standard of care required of a professional person providing a service.

If the Contractor engages a consultant to do the design work and that contract is under the NEC Professional Services Contract (PSC), the standard of care required in that contract is 'to use the skill and care normally used by professionals providing services similar to the services', which can be interpreted as equivalent to reasonable skill and care. Hence a contractor who enters into a contract under the ECC, which does not include option X15, should be aware that in engaging a consultant under the PSC for the design of the works, he may be carrying the risk of the difference between the two standards.

Although not occurring very often, it is possible that a defect (as defined in the contract) appears in spite of the fact that the Contractor has exercised reasonable skill and care. In such a case, the Contractor is obliged to correct the defect. However, because the Contractor is not liable for the defect, correction is a compensation event.

Should the Employer pay the full amount due for the work the Contractor has done, or should he temporarily keep back a proportion? Option X16: Retention

In most standard conditions of contract, there is provision for some deduction from the amounts otherwise due to be paid to the contractor. The amount deducted is usually paid back to the contractor in two stages – at completion and some time after completion. The historical reason for this is probably to provide some security for the Employer in the event that defects appear afterwards, or that the works are later found to be non-compliant in some way.

Retention is in addition to the security provided by the fact that payment for work occurs some time after the Contractor has carried it out. Without this option, retention is not deducted. The amount retained is often a stated percentage of the amount otherwise due, with a maximum retention expressed as a percentage of the original tender sum.

This option allows for a retention-free amount as stated in the contract. This eases the Contractor's cash-flow burden at the beginning of the contract, which is already considerable because of establishment costs and not receiving any payment for work done until, say, some two months after the start of the works. After the retention free amount is reached, the retention then consists of a percentage deduction. Repayment of the amount retained is similar to the provisions in most standard forms of contract.

What price should the Contractor pay if he fails to carry out his obligations? Option X17: Low performance damages

In certain types of project, works are designed and then built to a performance specification. After they have been finished, the works are subject to commissioning tests to check that they comply with the specified performance criteria. This is frequently the practice in projects that include mechanical and electrical work. If the commissioning tests reveal that the works do not comply with the specified standard, the options available to the Employer would be to require the Contractor to demolish the works and redesign and rebuild (not a practicable solution in most cases). Alternatively the parties could negotiate a reduction in payment to the Contractor for the breach of contract. Option X17 recognises the possibility of such an occurrence and allows for payment of damages at levels agreed in the contract, to reflect the reduced level in the performance achieved. Thus, the contract states a range of different performances, together with the damages payable by the Contractor for each level below a particular specified level. A minimum acceptable level is usually stated in the contract. Under English law, penalties are not enforceable. Hence the amount of damages stated in the contract should not be greater than a genuine estimate of the financial damage suffered by the Employer as a result of the reduced performance.

Should there be some financial limit to the Contractor's legal liability? Option X18: Limitation of liability

This option permits the fixing of limits to various liabilities to the Employer, which the Contractor may have under the contract, or even outside the contract in the case of his total liability. The Contractor's liability may be considerable, particularly in design and build contracts and where the Contractor may be working on or near to an asset belonging to the Employer. The potential liability in such cases may be out of all proportion to the value of the contract – hence the need, for practical and commercial reasons, to limit the Contractor's liability. The separate liabilities provided for are indirect or consequential loss, damage to the Employer's property and the consequences of defects due to the Contractor's design. In addition there is provision for limiting the Contractor's total liability for all matters. In many jurisdictions there is a statutory limit to the time after which the parties' obligations under a contract cease (the limitation period). Where there is no such statutory limit or where the parties wish to change the limitation period, the final clause in this option can be used to introduce an 'end of liability date'.

Should the Contractor be rewarded for meeting specified targets? Option X20: Key Performance Indicators

An important aspect of recent commercial and industrial life has been the introduction of targets and incentives with the object of achieving continuous improvement. The aim of Key Performance Indicators (KPIs) is to provide an incentive to the Contractor to achieve the Employer's objectives. This is done by deciding a method of measuring performance of different kinds, and comparing the actual performance achieved with previously established targets. A bonus is paid to the Contractor where he achieves the target. Since the purpose of this option is to motivate and encourage the Contractor to improve his performance, rather than apply penalties for failure to reach a target, there is no provision for penalties if he fails to meet a target.

The KPIs are listed in an 'incentive schedule', which includes how performance is measured and states the target, as well as the bonus to be paid to the Contractor on achieving the target. It is the Contractor's obligation to periodically measure his performance against each KPI and report the results to the Project Manager. Where the Contractor fails to meet a target he is required to submit proposals for improvement.

What is necessary to make the contract comply with UK legal requirements? Option Y(UK)2: The Housing Grants, Construction and Regeneration Act 1996

The construction part of this Act was introduced to deal with two matters, namely adjudication of disputes and payment. The first of these is dealt with by selecting the appropriate dispute resolution option W2. Payment is dealt with by incorporating this option Y(UK)2. It should be used only where the Act applies to the work in question. The Act (and this option) introduces two terms, namely a 'due date' and a 'final date' for payment. The due date is seven days after the assessment date, as defined in Section 5 of the core clauses. The final date is 14 days after the due date (or other period stated in the Contract Data). The Employer is required under the Act to give notice to the Contractor of the payment to be made and this notice takes the form of the Project Manager's certificate. If the Employer does not intend to pay the amount certified, he must provide a 'notice to withhold payment'. Failure to do so entitles the Contractor to suspend performance. If he exercises this right, he is entitled (under the Act) to an extension of time equal to the period during which payment was withheld. Additionally under Y(UK)2, he is entitled also to additional costs, as his proper exercising of his right to suspend constitutes a compensation event. The purpose of the Act is to provide a practical remedy for a contractor who is not paid the amount due at the appropriate time.

Should third parties have any rights even though they are not parties to the contract? Option Y(UK)3: The Contracts (Rights of Third Parties) Act 1999

This Act permits third parties to exercise rights under a contract even though they are not parties to that contract. The previous edition of the ECC eliminated third party rights – a provision that the Act allowed. However, under NEC3 provision has been made for specific rights to be enforced by people or organisations stated in the contract. This may be particularly useful where the parties wish to give certain rights to others, such as partners, under option X12.

Is any fine tuning of the conditions of contract required to meet the specific requirements of the Employer and the Contractor and to suit the circumstances of the project? Option Z: Additional conditions of contract

In spite of the flexibility available in the structure of the ECC, it will often be necessary to make additional provisions to suit the circumstances of a particular project – the fine tuning. These additional z clauses comprise the bespoke element of the contract and are usually decided by the Employer but sometimes negotiated between Employer and Contractor. Additional clauses should not be used to change substantially the overall balance of risk between the parties as this may cause conflict within the contract. Additional clauses should be carefully drafted to harmonise with the standard clauses avoiding overlap wherever possible.

The roles, duties and powers of the Employer, the Contractor, the Project Manager and the Supervisor are described in detail throughout the conditions of contract

The contracting parties are the Employer and the Contractor, each of which has specific powers and duties described in the contract. In addition, other parties are referred to. These are the Project Manager and Supervisor, both of whom are appointed by, and operate on behalf of, the Employer. Subcontractors are defined in the contract to distinguish them from suppliers. The Adjudicator is named in part one of the Contract Data. His function is to decide disputes referred to him by one of the contracting parties, under option W1 or W2. 'Others' is also a defined term. Although there are a number of references to design, designers as such are not expressly mentioned in the contract. Both the Employer and the Contractor will normally employ designers either as employees or as consultants in order to fulfil their respective design roles. In a contract that follows the traditional pattern of the Employer designing the works and the Contractor building, the Employer will need a designer to design the works. In the case of a design and build contract, the Contractor may be required to design a part or the whole of the works. In such a case, the Employer's designer will be needed to design that part of the works which is his responsibility or at least to specify what the Contractor is to design together with design criteria. In all cases the Contractor will need the services of a designer to design temporary works.

The roles, duties and powers of these various parties are described throughout the conditions. However, section 2 of the conditions covers the main responsibilities of the Contractor. The kind of people appointed to these positions will depend on a number of factors, the most important of which is the nature and extent of the work to be carried out. Thus, the person appointed as Project Manager for a building project may well be an architect, or, for an engineering project, a civil, mechanical or chemical engineer.

The basic duty of most of the above parties is stated in the first clause of the contract. Presumably it is placed at the beginning because it underpins the whole contract. It describes not only what the parties must do but also how they are to act. The first of these, requiring the parties to act as stated in the contract, is inherent in the nature of contracts. The various actions are generally described throughout the conditions in the present tense and the use of the word 'shall' makes the actions mandatory. The second duty expressly states that the parties are to act in a spirit of mutual trust and co-operation. It is not usual for standard contracts to state how parties are to carry out their duties, even though the duty of co-operation has sometimes been implied. This requirement follows the recommendation of the Latham Report and has been the subject of much legal comment, particularly in relation to how it can be enforced. But its inclusion in the ECC is largely pragmatic, for the benefit of the parties and in the interests of the project, with the aim of avoiding disputes and litigation.

There is very little mention of the Employer in the contract, but he has some important basic rights and responsibilities

The Project Manager and Supervisor exercise most of the day-to-day management and administrative duties and powers on behalf of the Employer. However, there are a number of basic matters that are reserved exclusively for the Employer in his role as one of the contracting parties. These include certain rights over the Contractor's design, taking over the works or using part of the works before they are completed, acquiring the title to plant and materials, insurance, termination and the right to refer disputes to adjudication.

The Contractor has many fundamental obligations

While the main responsibilities of the Contractor are described in section 2 of the conditions, many other responsibilities are contained in the rest of the conditions. His main obligation is to provide the works, and what and how he is to provide is covered in detail in the Works Information. Any parts of the works that the Contractor is to design are described also in the Works Information, and procedures for obtaining acceptance by the Project Manager of the design details are set out in section 2. As in most standard contracts, the Contractor is responsible for designing any necessary temporary works but the Project Manager is entitled to have details of these should he so wish.

Before the Contractor subcontracts any of his work, he is required to seek the acceptance of the Project Manager for the appointment of a Subcontractor and the conditions of the subcontract. This goes further than many standard conditions and illustrates the greater involvement of the Employer's Project Manager in the contract. The Project Manager's involvement is increased still further in the cost reimbursement options, under which the Contractor is required to submit the Contract Data of the subcontract.

The Project Manager fulfils a crucial role in the successful managing of the contract

He is appointed by the Employer and is named in the contract. His role is crucial in the management of the contract. He has extensive powers and duties in the contract and it is important that he has the requisite qualifications to carry these out. The extent to which the Employer may restrict or qualify these powers and duties is a matter between the Employer and the Project Manager. But the Contractor may assume that the Project Manager has the necessary powers to fulfil his role in administering the contract. The contract contains remedies for the Contractor in the event that the Project Manager fails to act within the times stated in the contract. Thus it is incumbent on both Employer and Project Manager to ensure that any restrictions on his authority do not prevent the Project Manager from functioning effectively.

  www.neccontract.com

The Project Manager should be a person with business experience capable of exercising commercial judgement, particularly in decisions concerning payments to the Contractor and in assessing the effects of change. On major projects it would be normal for the Project Manager to delegate some of his powers and duties to others who may be better qualified to fulfil parts of this role. The authority to delegate is unrestricted in the contract. On the other hand, for small contracts the Employer may appoint the same person as both Project Manager and Supervisor.

Another important role of the Project Manager is that of certifier. He is required to issue certificates in relation to such things as payment, taking over of part of the works, completion and termination. In all of these, skilled judgement is required.

The Supervisor's main role is to inspect and test the Contractor's work

Like the Project Manager, the Supervisor is appointed by the Employer, acts on his behalf and is named in the contract. He has discrete powers and duties in the contract mainly in relation to checking the Contractor's work. His nearest equivalent in traditional contracts is the inspector or clerk of works except that in the ECC he is responsible to the Employer, not to the Project Manager. He may delegate any of his powers and duties to others and this may be essential on large projects. As in the case of the Project Manager, there is no restriction on what he may delegate.

The main references in the conditions of contract to the Supervisor are found in section 4. These relate to inspection and testing of the Contractor's work and materials, the searching for defects and the formal notifying and correction of defects. This section also includes a mechanism for accepting defects by agreement. This is an extremely useful procedure in cases where, although technically work is non-compliant, its correction is not strictly necessary or even possible. However, the Supervisor's powers do not extend to accepting defects – this is solely the province of the Project Manager.

The Supervisor also has a certifying role but only in respect of the Defects Certificate. This is a statement of defects, if any, which the Contractor has not corrected by the 'defects date'.

The appointment of Subcontractors and certain terms of subcontracts are subject to acceptance by the Project Manager

The Contractor may subcontract parts of the works, but that does not affect in any way his basic responsibility to provide the works. Because of the need to distinguish between suppliers (not defined as such in the conditions) and subcontractors, the latter are defined in a specific way. There are three parts to the definition. The third part requires that to qualify as a subcontractor, the supplier of plant and materials for the works must have carried out some design that is specific in relation to the works. In other words, the definition excludes firms that supply standard, 'off-the-shelf' items.

The clauses dealing with subcontracting encourage the Contractor to use a NEC contract for the subcontract. Suitable NEC contracts would be the Engineering and Construction Subcontract (ECS) or the short subcontract (ECSS) or the Professional Services Contract (PSC). The last of these would be suitable for subcontracting design work to a designer. While use of a NEC contract is not mandatory, the benefits to the Contractor are obvious in that the two contracts are almost back to back, leaving little residual risk with the Contractor. The main option in the subcontract may well differ from that of the main contract and, similarly, the secondary options, for a number of reasons, may be different. But the requirement of 'mutual trust and co-operation' is effectively mandatory in the sense that its absence provides a valid ground for rejection of a subcontract proposal. The intention here is to clearly ensure that this fundamental condition extends down the whole of the supply chain.

There is no provision in the contract for nomination of subcontractors.

The Adjudicator's role is to decide disputes

The procedure specified in the contract for resolving disputes, is that of adjudication. Since its inclusion in the first edition of the NEC, the right to have disputes in a construction contract decided by an adjudicator has been enshrined in statute in the UK's Housing Grants, Construction and Regeneration Act 1996. In recent years the independent role of the architect in the JCT contracts or engineer in the ICE Conditions of Contract, in dealing with disputes, has been increasingly called into question. The appointment of a third party to fulfil this specific role should satisfy the critics. However, experience of adjudication has produced its own problems but these are of a different kind. Generally, the introduction of adjudication has made available a quick and relatively cheap means of resolving the majority of disputes that occur in construction.

The Adjudicator may be appointed by a number of different methods, but the objective should always be to appoint a person who has the confidence of, and is acceptable to, both parties. He should be appointed under a tripartite contract, using the NEC Adjudicator's Contract (AC). He is called upon only when the parties have failed to resolve a dispute. He is required to decide the dispute within 28 days (or longer agreed period) and is paid on a time basis. His costs under the AC are shared equally between the parties unless otherwise agreed.

The Adjudicator's decision is legally enforceable but a dissatisfied party may take the dispute further to the 'tribunal', which is specified in the contract as either (normally) arbitration or litigation.

The terminology used in the ECC is generally different from that in many traditional contracts

In the published literature no reasons have been given for this, but some possible reasons are given in the previous chapter. Some similar terms have, however, been used but often with different meanings. Some of the more important terms are described in the following.

Defects date This is a date that is a fixed period after Completion. The fixed period is stated in the contract and, following much tradition, is frequently stated as being 12 months. Its nearest equivalent in some traditional contracts is the end of the maintenance or defects liability period.

Defect This has a specific definition, which incorporates an element of Contractor's liability. It excludes defects for which the Contractor is not liable.

Defects Certificate This is a defined term and is different from the traditional maintenance certificate.

Completion This is a defined term that seeks to remove, as far as possible, subjective judgement in deciding when works have been completed.

Equipment This is restricted to anything of a temporary nature, which is removed from the site and working areas on or before completion of the works. It excludes permanent plant and equipment which become part of the works.

Working Areas This defined term includes the site and other areas used by the Contractor in providing the works.

Works Information This term is defined in the contract. The Works Information is contained in those documents stated in the Contract Data. These are very important documents in that they describe in detail what the Contractor is required to do. They would normally comprise mainly drawings and specifications.

Site Information This is another defined term that relates to the whole of the site and its surroundings.

Compensation events This is not a defined term but refers to events that, if and when they occur, may entitle the Contractor to additional payment (or in some cases reduced payment) and time extension. They are mainly listed in section 6 of the conditions together with the procedures for managing them and their consequences.

Activity Schedule This is a pricing document used in main options A and C. It determines the lump sum price for the works in the case of option A. It is a list of activities priced by the Contractor. In the case of option C it is used to determine the target (the Prices) from which is calculated the Contractor's share at the end of the contract.

Bill of Quantities This is a traditional term widely used for many years in construction. Its use in the ECC follows that in the ICE conditions rather than the JCT. Correction of quantities to reflect the work described in the Works Information, which the Contractor actually carries out, is not generally a compensation event. It therefore creates a remeasurement or admeasurement contract.

The Prices This is a term that has different definitions according to the main option used. It forms the basis on which payments to the Contractor are calculated.

The Price for Work Done to Date This term has different definitions in the main options. It is used to calculate the periodic payments made to the Contractor progressively throughout the contract period.

Defined Cost This term replaces the 'Actual Cost' in earlier editions of the ECC. It has different definitions in the main options. It has two main uses. The first is to assess the financial element of compensation events and the second is to determine payments to the Contractor in the cost reimbursement options C to F.

Disallowed Cost This is a defined term. It consists of elements of cost that are deducted in calculating the Defined Cost.

The Schedule of Cost Components To calculate Defined Cost it is necessary to list those elements of cost that are to be included. It is used only in options C, D and E.

The Shorter Schedule of Cost Components This is a shorter version of the Schedule of Cost Components, used for assessing compensation events in options A and B. It may also be used for this purpose in options C to E but only by agreement.

The ECC contains some provisions that are quite different from those in traditional contracts

Some of the more important and distinguishing features of the ECC are described in the following.

Communications Clause 13 recognises the importance of effective communication between all parties in managing the contract efficiently. It does not permit purely verbal communication as a valid means of communication under the contract; it requires that communications are in a form that can be 'read, copied and recorded'. This does not, of course, prevent parties communicating with each other verbally, but to have contractual effect an instruction or other communication must be in written form. The clause does not allow for confirmation by the Contractor of verbal instructions. The clause requires parties to reply to communications within definite periods of time. Failure by the Project Manager or Supervisor to reply within these times constitutes a compensation event.

Acceptance procedures are also covered in Clause 13. Throughout the conditions, whenever the Contractor is to make a submission to the Project Manager for his acceptance, the criteria for rejection of the submission are also stated. These criteria are stated in objective terms as far as possible. This does not prevent the Project Manager from rejecting a submission for other reasons, should he so wish, but in that case rejection is a compensation event. This minimising of purely subjective grounds for rejection of a Contractor's submission, is likely to reduce the risk of disputes.

Early warning This has been described by one construction lawyer as the jewel in the crown of the NEC. The procedure described in clause 16 gives priority to the parties co-operating in solving a problem as soon as possible after it arises. Procedures in many traditional contracts have the effect of motivating the parties to adopt a contractual and defensive stance when problems arise, with both parties reserving their position. This clause requires the Contractor formally to notify the Project Manager (and vice versa) of any matter that could affect the outcome of the contract. The parties are then encouraged to co-operate in seeking solutions to the problem that has been identified, normally by means of a meeting (now called a risk reduction meeting). The agenda for the meeting is set out in the clause, the objective of the meeting being to discuss the problem, find and agree solutions, and decide on the actions to be taken. Liabilities and payment under the contract will follow from the decisions of the meeting.

The programme The programme is a very important document in the ECC. Because of the range of options available, the Project Manager is much more involved in the Contractor's activities – not only in what he does but also in how he does it. Clause 31 allows for a tender programme (which becomes the first programme on acceptance of a tender) and also a programme submitted by the Contractor after the Contract Date. The importance of the programme is demonstrated by the long list of matters that are required to be included in the programme. Apart from various dates which are fundamental in any programme, the list includes statements of methods that the Contractor proposes to use to provide the works together with the resources he plans to use. It also includes time allowances for matters that are at the Contractor's risk, and timing of the work of Others. Such detail is of considerable benefit to both parties when it becomes necessary to assess the impact of change. The detailed information on the programme, particularly that relating to the contractor's methods, float and time risk allowances is likely to be very useful in assessing compensation events. While the programme is no more than a statement of the Contractor's intentions and proposals, certain dates on the programme acquire contractual status. This is clear from the list of compensation events in section 6 of the conditions, three of which refer to dates on the Accepted Programme. One of these compensation events also makes clear that in the event of one of the 'Others' not performing in accordance with the programme, that risk is taken by the Employer.

Take over The Employer has authority to take over and use a part of the works (clause 35) before it has been completed. While such a provision is of considerable benefit to the Employer, the Contractor may have to complete the outstanding work with the Employer in occupation, at extra cost. Early take over is therefore included in the list of compensation events. Completion will occur later and the completion certificate will be issued independently of take over.

Managing Defects As soon as Defects are identified they are formally notified (Clause 42) and then corrected (Clause 43) within stated periods of time (the defect correction period). However, occasionally, conditions may be such that it becomes impossible to correct defective work. Circumstances may occur also where the cost of correcting a Defect may be great and its correction may be technically, though not contractually, unnecessary. Clause 44 provides a mechanism to deal with a situation that is not uncommon in many contracts. Under this clause, a Defect can be accepted by agreement of the parties, with some reduction in the Prices and possibly an earlier Completion Date.

Compensation events These are events that may entitle the Contractor to extra payment or, in some cases, reduced payment. They are thus matters that are at the financial risk of the Employer. They are listed in section 6 of the conditions. Additional compensation events are included in some of the main and secondary options. The procedures for notifying them are clearly stated in clause 61 and assessment of their financial and time effects are detailed in clause 63. In general, all compensation events are assessed as a package of time and money although, in some cases, if the operation affected by the compensation event is not on the critical path, the effect on the Completion Date may be zero. However, there is considerable flexibility provided in clause 62.1 whereby the Contractor may submit a quotation (an alternative quotation) comprising only the cost element but no time extension. The cost of the extra resources required to avoid delay to completion often represents good value to an Employer anxious to complete a project as soon as possible.

The general rule for extending the time for completion is clearly laid down in clause 63. This assumes extrapolation of planned resources for an activity to carry out any extra or changed work.

The financial aspect of a quotation for a compensation event is calculated by assessing the impact of the event on the Contractor's costs (Defined Cost), rather than by using rates and prices stated in the contract whether in the Activity Schedule or the Bill of Quantities. However rates in the Bill of Quantities under options B and D may be used but only if the parties agree.

When to use the ECC

From the above description it is clear that the ECC can be used for all kinds of construction projects required by a promoter or client (the Employer) to be built and perhaps designed by a Contractor. A separate form of contract is not required where the design of some or the whole of the permanent works is to be done by the Contractor; the ECC allows for any design by the Contractor. Proper management of projects inevitably means that the procedures stated in the contract should be meticulously followed.

There is clearly a size of project below which it would not be practicable to use the ECC. In such circumstances the short contract (ECSC) may be more suitable. The dividing line between the two depends more on such matters as complexity, risks, nature of the work required and ease of management, rather than estimated value in purely financial terms. The choice between the two is a matter of judgement in each case.

The ECC should not be used for the provision of professional services (where the NEC Professional Services Contract would be appropriate), for subcontract work (where the NEC Engineering and Construction Subcontract would be appropriate) or for maintenance-type contracts over a fixed period of time (where the NEC Term Service Contract would be more appropriate).

How to use the ECC

Having decided to use the ECC, the Employer's next task is to decide his contract strategy and, in particular, which main option he proposes to use to suit the circumstances of the project. He will then need to decide which secondary options to incorporate in the contract and draft any z clauses he wishes to add. One of his most time-consuming tasks is to prepare the Works Information. The effort and resources required for this task will largely depend on the design work, if any, which he requires the Contractor to carry out. He should then complete part one of the Contract Data and any further documents such as the Site Information, which are referred to in the Contract Data.

The next stage will depend on whether the Employer intends to invite competitive tenders, in which case he will prepare a list of tenderers and instructions for tendering. For this purpose, the tender documents will include part two of the Contract Data as a pro-forma, which tenderers will be invited to complete. Guidance on assessing the financial aspect of tenders is provided in the appendices to the published guidance notes. If the Employer already has some form of framework agreement with a number of contractors, he will invite one of them to submit a proposal and negotiate the terms in accordance with that agreement.

Once a contractor has been appointed, one of the first tasks of the Employer's Project Manager and Supervisor is to decide which powers he wishes to delegate and to whom. It is also usual practice to meet the Contractor to discuss the practical details of the contract procedures. An important aspect at this early stage is to develop a team building strategy to motivate all parties to co-operate in achieving the project objectives.

Exercises

(1) In the UK a 10 km-long single carriageway by-pass is to be built over difficult terrain to by-pass a small town. In addition to the earthworks and road works, it includes:

- Construction of a new sewage treatment works to replace the existing works which is on the line of the new by-pass (£1.5 m).
- A bridge crossing over a large river (£5.0 m).
- A soft ground tunnel 1.0 km long.
- Diversion of a 1050 mm dia high pressure gas main.

The employer has decided the general alignment of the new road and he has also decided to carry out the work under separate ECC contracts. As adviser to the employer, you are required to write a report on contract strategy including your recommendations for:

- Who is to design the various parts containing a design element.
- The number and content of the contracts.
- The main and secondary options for each contract.
- Co-ordinating and managing the contracts.

Make any necessary assumptions.

(2) (a) Compare and contrast the role of the ECC's Project Manager with that of the Engineer under the ICE Conditions of Contract or the Architect under the JCT contracts.

(b) Compare and contrast the role of the ECC's Supervisor with that of the Engineer's Representative under the ICE Conditions of Contract or the Clerk of Works under the JCT contracts.

(3) Compare the ECC's activity schedule as a payment mechanism with that of the bill of quantities. How would you decide which to use?

(4) Describe the status and function of the programme in the ECC and compare this with those of programmes in other standard forms of contract.

(5) Under a ECC contract, the contractor has completed the construction of a 600 mm dia pipe culvert under a busy railway line. Soon after it is completed, the supervisor checks and finds that it is 3 m out of position, The cost of removing it and constructing it in the correct position would be high and would cause major delay and disruption. However, it is agreed that the actual location is satisfactory from a design point of view subject to minor realignment of the water course at the inlet and outlet. The supervisor consults the railway authorities and finds that they are not concerned provided the legal situation and plans are corrected. Describe the procedure that is available to deal with this situation?

(6) Discuss the merits of partnering in construction.

3 The Engineering and Construction Subcontract (ECS)

Origins

When the consultative and first editions of the New Engineering Contract (as it was then called) were published, a subcontract version of each was published at the same time. This practice has been followed with the second and third editions. Thus when Sir Michael Latham published his report 'Constructing the Team' in 1994, with its recommendation for a wholly inter-related package of documents, standard NEC subcontract forms were already available.

The ECS is essentially back-to-back with the ECC

The benefits of back-to-back subcontract conditions are considerable especially for the Contractor. Where main and subcontract conditions are different, the Contractor's task of interpreting the different conditions, often using different terminology, may be very complex. In these circumstances, careful examination of both contracts is essential in order to identify risks and liabilities that the Contractor may not be able to pass to the Subcontractor.

However, the ECS is not wholly back-to-back with the ECC. The most important area where this may be the case is in the selected options for each contract. If, for instance, the main contract incorporates the target main option C, it is likely that most of the subcontracts will not be based on option C. These sub-contracts are more likely to use main option A or B. Similarly, not all the main contract secondary options may be used in the subcontract. For instance, in a major construction contract extending over a number of years, the Employer may have elected to take most of the inflation risk by incorporating option X1. A contractor under a subcontract may, however reasonably, require the sub-contractor to take the inflation risk where the subcontract is for a relatively short period.

The ECS, unlike many standard forms of subcontract, is not dependent on, and does not incorporate by reference, any of the terms of the main contract. It is a 'stand-alone' contract, except that factual information regarding the identity of such parties in the main contract as the Employer, Project Manager, Supervisor and Adjudicator are included in the subcontract documents. The Contractor will normally prepare the Subcontract Works Information. Much of this may be information reproduced from the Works Information in the main contract. But other information concerning matters only between the Contractor and Sub-contractor, will also be included. Hence, to this extent, the contracts are not back-to-back.

The management procedures in the ECC are reproduced in the ECS for obvious reasons. However, the time periods in the subcontract have been adjusted to allow for the passage of information down to and up from the Subcontractor.

The introduction of key dates in the NEC3 contracts is particularly relevant to subcontract work. It facilitates the Contractor's management of the interfaces (both physical and time) between Subcontractors, and between Subcontractors and the main contractor. In all cases the subcontract programmes should relate closely to the main contract programme which becomes, in effect, a master programme.

Provisions for subcontracting in the ECC allow the Project Manager to exercise a certain degree of control

A Subcontractor is defined in the list of definitions in the ECC. The main purpose of the definition is to distinguish a subcontractor from a supplier. Before the Contractor appoints a subcontractor, he is required to obtain the acceptance of the Project Manager. He is also required to submit to the Project Manager for acceptance the conditions of contract that he proposes to use for the subcontract. There are some exceptions to the latter requirement, namely those subcontracts using one of the NEC family of contracts or where the Project Manager does not wish to have a submission.

The cost reimbursement and target options include the additional requirement that the Contractor may have to submit the contract data for a subcontract if requested by the Project Manager. The reason for this lies in the cost reimbursable nature of these contracts and the need to exercise some control of costs on behalf of the Employer. A similar clause is included in the ECC management option F, but this is also likely to include strict criteria in respect of procurement procedures. These have not been included as standard clauses as each management contract is likely to have different requirements, but they would normally be included in the bespoke z clauses.

It is a characteristic of the NEC family of contracts that whenever a party has to make a submission to the Project Manager or his equivalent, the criteria entitling the Project Manager to reject the submission are also stated. This does not prevent the Project Manager from rejecting a submission for reasons not listed in the contract, but rejection in these circumstances constitutes a compensation event entitling the Contractor to additional costs and possibly time. This may be relevant to the appointment of subcontractors, where the Employer may have some good reason (other than the ones stated in the non-acceptance criteria) for not accepting a particular subcontractor.

There is no provision for nominated subcontractors None of the NEC family of contracts make provision for the nomination by the Employer of particular subcontractors. For a number of reasons the trend in the UK in recent years has been to avoid nominated subcontracting. The published guidance notes of the ECC explain why this policy has been followed and suggest alternative ways of dealing with the matter. These include making the Contractor responsible for placing work with domestic subcontractors of his choice, dividing the work into separate main contracts, and including lists of acceptable subcontractors.

Essential and distinguishing features of the ECS

Since many of the clauses of the ECS are identical or similar to those in the ECC, the comments in the chapter on the ECC to which the reader is referred, are not repeated.

The published documents As in the case of the ECC, preparation of tender and contract documents for a subcontract requires considerable thought and effort if a satisfactory outcome is to be achieved. Choices have to be made, usually by the Contractor, to suit the particular circumstances in each case. A subcontract under the ECS will comprise:

(a) Core clauses. These must be included in every subcontract.
(b) One main option selected from the five main options A to E. Selection of the main option determines the allocation of risk between the parties and how the Subcontractor is paid.
(c) A dispute resolution option W1 or W2, according as to whether the UK's Housing Grants, Construction and Regeneration Act 1996 applies to the subcontract work.
(d) Secondary options selected from the numbered options X1 to X20 (though some numbers have not been used in the ECS) and the two options numbered Y(UK)2 and Y(UK)3, which relate to work in the UK. This selection determines the allocation of further risks.

(e) Additional conditions of contract indicated by the letter z. These are normally decided by the Contractor or by agreement between the parties. Because of the options available, the need for additional conditions in the form of z clauses should be minimal.

(f) Subcontract Data. This is in two parts, part 1 being prepared by the Contractor and part 2 prepared in part by the Contractor for completion by tenderers or the selected subcontractor. The Subcontract Data contains information specific to the subcontract. Some information in the Subcontract Data takes the form of reference to other documents, which are thereby incorporated into the subcontract.

One of the most important decisions the Contractor has to make is to decide which main option to use

There are five main options: A, B, C, D and E.

The Subcontractor carries the greatest risk under options A and B and the least risk under option E. Selection is determined by the nature of the subcontract work to be carried out, and how the risks are to be dealt with. Placing maximum risk on the Subcontractor may not always be in the best interests of the Contractor or of the project.

The lump sum option. Option A: Priced subcontract with Activity Schedule

This option is fully described in the chapter on the ECC. The Contractor, of course, takes the place of the Employer, Project Manager and Supervisor. Because of its ease of administration, it is likely that this option will be widely used for many subcontracts where the subcontract work can be clearly specified.

The remeasurement option. Option B: Priced subcontract with Bill of Quantities

This option has been widely used in the UK for many years as a basis for both main and subcontracts. Because it is a remeasurement contract, the Contractor carries the risk of changes in quantities as between the Contractor and Subcontractor. Its use is appropriate where it is likely that the quantities will change to any great extent. It is most likely to be used when the main contract is under option B or D.

The target option with the target established as a lump sum. Option C: Target subcontract with Activity Schedule

Because of the greater administration involved in a target subcontract, this option is likely to be used only for major subcontracts where the Contractor wishes to motivate the Subcontractor to keep costs to a minimum. It could be used where the main contract is carried out under the lump sum option A, but for a specific part of the works which is high risk, the Contractor may prefer to use this target subcontract in order to maximise his return and minimise his risk.

The target option with the target established as a remeasurable sum. Option D: Target subcontract with Bill of Quantities

This option is very similar to option C except that the target price is adjusted for remeasurement of quantities as well as for compensation events. It should be used in subcontracts in preference to option C where quantities of work in the subcontract works are likely to change substantially.

The option where the Subcontractor is paid his costs. Option E: Cost reimbursable subcontract

Under this option there is little incentive to motivate the Subcontractor to minimise costs and hence the option is not likely to be often used for subcontract work. It may, of course, be used where the main contract is also being carried out under option E and in circumstances of emergency or high-risk work. If it is used where the main contract is under one of the priced options, A or B the Contractor carries the risk of the quality of performance of the Subcontractor.

The dispute resolution options determine the procedures for dealing with disputes

The dispute resolution options W1 and W2

The basis on which the choice between the two options is made is described in the chapter on the ECC. As far as the ECS is concerned, the clauses are similar to those in the ECC except for the introduction of extra procedures in the ECS to deal with 'joinder' of disputes.

The Act defines construction contracts (to which the Act applies) in terms of construction operations. These are defined in both positive and negative terms. Thus it is possible that, in a particular contract, there may be a combination of both construction and non-construction operations as defined in the Act. The Act states that in these cases the Act applies only to construction operations. However, in practical terms only one dispute resolution procedure would be included in a contract. Thus, if option W2 is selected, the dispute resolution procedures would apply to the whole contract. This is relevant to the ECS in that even if the subcontract works are not construction operations as defined in the Act, it would be advisable to select W2 as the appropriate dispute resolution procedure. The advantage of this is that it ensures back-to-back procedures for both main and subcontracts, and facilitates the use of the joinder clauses in both contracts.

The procedures in both options W1 and W2 allow for a dispute that is common to both contracts to be decided as between the three parties by a single adjudicator – the named adjudicator in the main contract. A similar procedure is contained in the ECS for disputes between the Subcontractor and a Subsubcontractor to be decided by the subcontract adjudicator.

The secondary options

Various further risks in carrying out the subcontract works need to be considered to decide how they can be reduced and whether they should be borne by the Contractor or the Subcontractor. This is the function of the secondary options

The number of secondary options in the ECS is the same as that in the ECC. The contents of the clauses are similar to those in the ECC except for changes to reflect the different names of the parties involved. For comments on each secondary option, reference should be made to the chapter on the ECC. Comments specifically relevant to subcontracts are given in the following.

Should the risk of price increases be carried by the Contractor or the Subcontractor? Option X1: Price adjustment for inflation

Whether or not to include this option is determined by whether it is included in the main contract. However, it is possible that a main contract of long duration and which includes option X1, may involve a number of subcontracts of relatively short duration. In this case it may be reasonable to omit option X1 from some of the subcontracts and require the subcontractors to allow in their tendered prices for price rises during the period of the subcontract.

Is the Subcontractor to be paid in more than one currency? Option X3: Multiple currencies

In many cases where the main contract includes option X3, it may not be required in the subcontracts where the Subcontractor is to be paid in a single currency.

What security does the Contractor require in the event of the Subcontractor's failure to carry out his obligations? Option X4: Parent company guarantee

Many contractors in the construction industry have subsidiary companies specialising in particular types of work. This option may, therefore, be relevant where such specialist companies are employed as subcontractors in that it gives the Contractor greater security in the form of a guarantee by the parent company.

Does the Contractor require sections of the subcontract works to be completed at different times? Option X5: Sectional Completion

There are two reasons why a contractor may wish to include this option. First, where the main contract specifies sectional completion and the relevant work, or at least part of it, is to be carried out by a subcontractor, the Contractor will require this obligation to be passed to the Subcontractor with appropriate incentives for early completion as appropriate, or damages for failure to complete in time. Second, the main Contractor would include this option when he wishes the Subcontractor to complete sections of work at specific times to enable the Contractor to adhere to his own programme.

Does the Contractor want the subcontract works to be completed as soon as possible? Option X6: Bonus for early Completion

The inclusion of this option may reflect the Contractor's incentive in the main contract to complete as soon as possible, or to help the Contractor to adhere to, or even improve upon his own programme.

What happens when the Subcontractor fails to complete the subcontract works on time? Option X7: Delay damages

The amount of delay damages should not exceed a genuine estimate of the damage suffered by the Contractor in the event of late completion. It is possible that such a failure by a subcontractor to complete on time results in delay to the main contractor, the financial effect of which is out of all proportion to the value of the subcontract. In these circumstances a subcontractor may insist on a lower level of damages and, if the amount is agreed between the parties, the lower level and any limit to the damages should be made clear in the Sub-contract Data. Some contractors may prefer not to include this option and rely, instead, on recovering unliquidated damages, or damages at large.

How can the Subcontractor be encouraged to work with other parties as a team, on the project, to produce a successful outcome? Option X12: Partnering

The system of partnering under the NEC is described in the published guidance notes and in the chapter on the ECC. NEC policy is to encourage co-operation of all parties in the supply chain. Thus, subcontractors would normally be included in partnering arrangements for a project and major subcontractors may even be included in the core group of partners.

How can the Contractor obtain some financial security when the Subcontractor fails to perform? Option X13: Performance bond

Judging from the number of disputes that arise in subcontracts, compared with those in main contracts, the need for security of a subcontractor's performance and, hence, the need for including this option, may be greater in subcontracts.

How can the Contractor help to finance the subcontract during the early months? Option X14: Advance payment to the Subcontractor

Inclusion of this option is a matter of judgement by the Contractor. If the option is included in the main contract, he must assess whether he wishes to pass on the benefits (or a part of them as appropriate to the subcontract) to the Subcontractor. If he decides not to include an advanced payment, the Subcontractor will have to bear an increased financing burden, which will be reflected in higher subcontract prices.

Should there be some limit to a Subcontractor's legal liability for his design work? Option X15: Limitation of the Subcontractor's liability for his design to reasonable skill and care

If this option is included in the main contract, there is little reason why a contractor should insist on a higher standard for design carried out by a subcontractor. If this option is not included in the main contract, it is possible that a subcontractor will insist on its inclusion in a subcontract that includes design work. In these circumstances, the Contractor carries the risk of the difference between the two standards of care.

Should the Contractor pay the full amount due for work that the Subcontractor has done, or should he keep back a portion? Option X16: Retention

It has been common practice in the UK, for many years, to retain a proportion of the money due to a contractor or subcontractor. Retention under the ECS can only be applied if this option is included. In the normal case where the option is included in the main contract it would also normally be included in subcontracts. However, where subcontract works are due to be completed before the main contract works, payment of the retained amounts may have to be made by the Contractor before he receives payment by the Employer.

What price should the Subcontractor pay if he fails to carry out his obligations? Option X17: Low performance damages

This option would normally be included in a subcontract for work which is subject to low performance damages in the main contract.

Should there be some financial limit to the Subcontractor's legal liability? Option X18: Limitation of liability

For practical and commercial reasons, the need for limiting a subcontractor's legal liability may be much greater than in the case of the Contractor's liability in the main contract. The extent of these limitations is a matter for the Contractor to decide in the first instance but, probably, also limitation of liability may be the subject of negotiations with potential subcontractors.

Should a Subcontractor be rewarded for meeting specified targets? Option X20: Key Performance Indicators

Where the main contract includes KPIs, there is no reason why back-to-back provisions should not be included in the relevant subcontracts. This ensures that the incentives to perform well are passed down the supply chain.

What is necessary to make the subcontract comply with UK legal requirements? Option Y(UK)2: Payments under the Housing Grants, Construction and Regeneration Act 1996

The comments under this heading in the chapter on the ECC also apply to the ECS. However, it is possible that particular subcontract works do not have to comply with the Act as they do not come under the definition of construction operations as described in the Act. But the exclusion of this option would deny the right of a subcontractor to suspend performance in the event of non-payment. To this extent, the subcontract would not be back-to-back with a main contract that was subject to the Act. The chief mischief, which the payment section of the Act is designed to deal with, is the non-payment for work, particularly in subcontracting. One of the curious results of the definition of construction operations in the Act is that a contractor may have suspension rights in the main contract but similar rights may not apply in a subcontract. Clearly, back-to-back provisions are desirable if disputes are to be avoided.

Should third parties have any rights even though they are not parties to the subcontract? Option Y(UK)3: The Contracts (Rights of Third Parties) Act 1999

The Contractor needs to give careful consideration to this question. Any rights given to third parties in the main contract would normally be given to them in a subcontract where applicable.

Is any fine tuning required to meet the requirements of Contractor and Subcontractor to suit the circumstances of the subcontract works? Option Z: Additional conditions of subcontract

In the interests of the project as a whole, additional clauses in a subcontract should, as far as possible, preserve the balance of risk in the main contract. As a result of their greater commercial strength, contractors may often be tempted to place onerous conditions on subcontractors through the medium of the z clauses. The temptation should be resisted.

The roles, duties and powers of the Contractor and Subcontractor are described in detail throughout the conditions of contract

The contracting parties are the Contractor and the Subcontractor, each of which has specific duties in the contract. In the ECS there are no parties equivalent to the Project Manager and Supervisor of the ECC. The Contractor may delegate any of his powers and duties in the same way as the Project Manager and Supervisor may in the ECC. The Subcontractor may subcontract some of his work to others who are defined as Subsubcontractors, but this is subject to acceptance by the Contractor. The Contractor is involved in the conditions of contract of any subsubcontract and in the target and cost reimbursement options the contract data are subject to the Contractor's acceptance. The Adjudicator is named in part 1 of the Subcontract Data and his function is to decide disputes that arise between the Contractor and Subcontractor. The Adjudicator may not be the same person as the person named in the main contract, but there are obvious advantages if he is the same person. If the subcontract includes a design element, the Subcontractor will require the services of a designer – either in-house or employed externally. As in the ECC, designers are not expressly mentioned in the ECS.

The roles duties and powers of the Contractor and Subcontractor are described throughout the conditions, though the Subcontractor's main responsibilities are described in section 2. The ECS includes few references to the Employer of the main contract, although he is identified in the Contract Data.

The Subcontractor has many fundamental obligations in the subcontract

The Subcontractor's main obligation is to provide the subcontract works. Any design by the Subcontractor is subject to acceptance by the Contractor, who in turn will need to follow similar procedures in the main contract to gain acceptance by the Project Manager. Much of the information provided by the Subcontractor in his programme, such as method statements, resources and work of others, will need to be incorporated in the Contractor's programme submitted to the Project Manager. This may require revision of the Contractor's programme in the main contract.

Under the section on Testing and Defects, there is a mechanism to enable the Contractor to accept defects by agreement and negotiation with the

Subcontractor. This is similar to the procedures in the ECC but, before agreeing to accept a defect, it would be expedient for the Contractor to obtain the Project Manager's acceptance under the main contract.

The ECS now has time-barring sanctions for failure to notify compensation events and submit quotations. The first of these has a rather shorter time of seven weeks (rather than the ECC's eight weeks) for notifying compensation events and thus it is expedient that the Subcontractor ensures his compliance with the timing provisions in the subcontract.

When to use the ECS

Under the provisions of the ECC, the Contractor is not obliged to use the ECS for subcontract work. For many small-value subcontracts, or those that are straight-forward in terms of risk and management, contractors may elect not to use the ECS. But the use of the ECS for substantial subcontract work has considerable advantages particularly in the back-to-back nature of its provisions.

How to use the ECS

Having decided to use the ECS, the Contractor's next task is to decide his contract strategy and, in particular, which main option he proposes to use to suit the circumstances of the subcontract work, and the terms of the main contract. He will then need to decide which secondary options to incorporate in the subcontract and draft any z clauses he wishes to add. The subcontract Works Information will comprise the relevant parts of the main contract Works Information together with other matters that the Contractor wishes to include. Details of any design work in the subcontract should be included. Most of the subcontract Site Information will be extracted from the Site Information provided under the main contract.

If the Contractor intends to invite tenders from a number of subcontractors, he will prepare part 2 of the subcontract data to be completed by tenderers. Assessing the financial aspects of subcontract tenders fairly should be such that all tendered information is brought into the competition. The guidance on assessing ECC tenders in the Appendices to the published guidance notes applies also to ECS tenders.

Exercises

(1) A contractor has been employed under the ECC using option C to design and build a factory for making yoghurts. This involves the design and installation of sophisticated plant and equipment of various kinds. Most of the work is to be done by subcontract.

Decide the main option you would recommend to the Contractor for the following subcontracts using the ECS, making any necessary assumptions:

- A new roundabout and other road works adjacent to the site to comply with the highway authority's requirements for access to the factory.
- Site clearance and earthworks.
- Piling for the main factory building.
- Structural steel framework for the building.
- Cladding and finishes for the building.
- Design and installation of refrigeration and other specialist equipment.
- Pavement works for the building surrounds and car and lorry parks.
- Landscaping.

4 The Engineering and Construction Short Contract (ECSC)

Origins

The first edition of the ECSC was published in 1999, some six years after the first edition of the ECC appeared. One of the claimed characteristics of the ECC is its clarity and simplicity, and thus the question arises, 'Why is a short version of the contract necessary?'. Strictly, anything that occurs on a large contract may also occur on a small contract. Hence, what is the justification for reducing or abbreviating the ECC to produce a shorter version? That may explain why the short form of contract was not produced until some years after the main ECC and the reason appears to be purely pragmatic. Thus, by implication, the short contract is suitable for contracts involving repetitive work, in situations where the employer may not be familiar with the NEC family of contracts, and where the parties are seeking clear and simple procedures both in preparing the contract documents and in managing the contract. The majority of construction contracts are of relatively small value and the detailed provisions and pro-cedures of the ECC cannot normally be justified for these smaller contracts.

The published information on the ECSC indicates that it is designed for use where the work and management techniques required are straightforward, and where risks are low. It is significant that its use has not been described in terms of the monetary value of the works. This recognises that the two are not related; a contract of small value may be complex and involve a high level of risk and, conversely, a contract of high value may be simple and straightforward in both risk and management terms, often involving work that is repetitive.

The second edition of the ECSC was published in 2005, at the same time as the other contracts in the family of NEC contracts. It incorporates most of the relevant amendments introduced in the updating of the other forms.

The ECSC is structured in a way that simplifies the preparation of the contract documents

At the beginning of the document there are several forms which, when fully completed and together with the standard clauses, create a binding contract. Thus the contract-specific details occur at the beginning of the document rather than at the end, as in the contract data of the ECC. There are no main options or secondary options as in the ECC. Flexibility of providing for different payment methods is limited in the ECSC in that there are no target or cost reim-bursement options. But the Price List provides a mechanism for paying the Contractor on either a lump sum basis or remeasurement basis or a combina-tion of the two. There is provision, also, in the contract data for additional condi-tions if an Employer wishes to incorporate some of the ECC secondary options. It would be necessary in such a case to amend the clauses of the secondary options as necessary to suit the ECSC.

Completion of the forms creates a contract

After the first title sheet, all the forms are described under the heading 'Contract Data'. The first two of these give details of those terms that appear in italics in the conditions of contract. Most of these are similar to those in the contract data of the ECC but there are some differences. The latter reflect the nature of the shorter version of the contract. For instance, delay damages are a standard requirement in the ECSC – not optional as in the ECC. Similarly, retention is standard rather than optional unless, of course, the Employer enters 'nil' in the contract data. As in the case of additional z clauses in the ECC, it is advisable to keep additional conditions to a minimum. The temptation to change substantially the balance of risk between the parties by means of additional clauses should be resisted.

Completion of the forms may be carried out manually or in the digital format that has been made available by the publishers.

The Contractor's Offer and the Employer's Acceptance

To decide when a contract is created, it has been common practice to identify an offer by one party and acceptance of that offer by the other party. Thus, the first form with the subheading 'Contractor's Offer' is to be completed by the Contractor. It includes two percentages that are to be tendered by the Contractor; these are needed solely for assessing compensation events. The first of these percentages is applied to the 'people' element of cost and the second is applied to the other elements. The percentages should be tendered at such a level as to include for costs not included in the 'Defined Cost', and also for overheads and profit. The offered total of the Prices (entered from the Price List) is also included but the statement of offer makes it clear that the amount to be paid to the Contractor is determined by the conditions of contract. Thus, in most cases, the total of the prices will be subject to change by remeasurement, compensation events and other matters.

The Price List is used to determine the payments due to the Contractor

The columns of the Price List are similar to those of a traditional bill of quantities. However, there is no method of measurement to determine what the items should be and how they are to be measured. But it is possible that the Employer may wish to specify rules under which a tenderer is required to draft and price the Price List. This means that, although the quantity of an item may have to be corrected, the number of items will not change other than adding items as a result of compensation events.

The Price List may be compiled by either the Employer or Contractor but the pricing is always done by the Contractor. It consists of lump sum items and remeasurable items. A lump sum item is only paid for when that item of work has been completed – there is no provision for payment for a proportion of an item. On the other hand, the Contractor is paid for remeasurable items according to the quantity of work he has completed.

The Works Information is a detailed statement of what is to be provided by the Contractor and how, and what things and facilities, if any, are to be provided by the Employer

Careful drafting of the Works Information is important if disputes are to be avoided. It consists of six parts. Most of this information is provided in the form of drawings and specifications. Where the Employer is clear on exactly what he requires, he will prepare these documents himself or obtain professional help to do so. In some cases the Employer may invite tenderers to offer certain details to a greater or lesser extent, to ensure the Works Information is complete. His choice of contractor may then depend on what each tenderer is offering as well as on price. The Employer may wish to apply some constraints on the Contractor in carrying out the works and these are to be listed in the section allocated for this purpose. Examples of constraints are hours of working, types of equipment and employment of local labour. Programme requirements are to be stated under a separate heading of the Works Information. These are likely to depend on the nature of the works and the extent to which the Employer is likely to be involved in the Contractor's activities. Where many changes are likely, the more information there is available in the programme, the easier it is to assess and manage the impact of the changes. By way of contrast, in small, simple contracts, all that may be needed is a start and completion date rather than a programme.

The simplest type of programme for use with the ECSC is the bar chart, sometimes known as a Gantt chart. But the Employer may wish to have details of methods of construction, equipment to be used and resources likely to be required. The detailed list of requirements of the programme in the ECC may be of value in deciding what should be included in the Contractor's programme for a particular contract under the ECSC. The final section of the Works Information is allocated to services and other things to be provided by the Employer. Again, careful drafting of this section is vital. For instance, the Employer may undertake to provide electricity, water, items of equipment and welfare facilities, in which case the precise nature location and quantity of these should be made clear.

Making available as much information about the site of the works as possible, is likely to reduce risks of the unknown and resulting disputes

The final section of the contract data is allocated to Site Information. This may take the form of nature of the ground, borehole data, soil test results, access to the site, boundaries of the site and details of adjacent buildings. Even though one of the compensation events in the ECSC relates to unforeseeable physical conditions, many employers disclaim responsibility for the Site Information with a view to transferring such risk to the contractor. In most circumstances, this transfer of risk is much more likely to produce disputes. 'Fairness' dictates that those who provide information should be responsible for it.

There is no Project Manager or Supervisor in the ECSC but the Employer may delegate authority to others

The contracting parties are the Employer and the Contractor, each of which has specific powers and duties described in the contract. In many cases the Employer may wish to appoint a person to act on his behalf in exercising his powers and undertaking his duties under the contract. The contract provides for this delegation of powers and duties, provided the Employer notifies the Contractor. Thus, one of the most important actions by the Employer at the start of any contract is to make clear who is acting on his behalf and what powers he has. As in the ECC, there is no similar provision for delegation by the Contractor. In practice, the Contractor will appoint an agent to act on his behalf. Because of its nature, administration of an ECSC contract should be much simpler than that of an ECC contract. Probably the area in which the Employer is required to carry out most administrative duties is in managing change. Thus, the procedures for notifying and assessing compensation events are important; they are very similar to those in the ECC.

The only other party named in the contract data is the Adjudicator. His function is to decide disputes referred to him by one of the contracting parties. It is important that the person appointed to this position is acceptable to both parties. No method of appointing an adjudicator is stated but there is a procedure for appointing a replacement in the event that the appointed person is not able to act. Many employers insert 'to be appointed' in the contract data, presumably in the expectation that there will be no disputes that the parties themselves cannot resolve, and thus no need for an adjudicator. But if the Employer wishes to name an adjudicator in the contract data, he may offer several names to tenderers and invite them to select one of them. Alternatively, the Employer may invite a name or a number of names from tenderers and select one of them. The advantage of naming the adjudicator in the contract is that it simplifies the resolution of disputes should they occur.

The Employer may use a person on his own staff to design the works or appoint a consultant to do so. Similarly, the Contractor may need to carry out design of temporary or permanent works or both. However, as in the ECC, designers as separate parties are not mentioned in the contract.

The precise roles of the Employer and Contractor are described throughout the conditions of contract. The main responsibilities of the Contractor are described in section 2. However, the principal obligation of both Employer and Contractor is stated in the first clause of the contract. This obligation underpins the whole contract. It describes not only what the parties must do but how they are to do it. It is included following a recommendation of the Latham Report and, in spite of being the subject of much legal comment and criticism as to its interpretation and enforceability, has been included in other NEC contracts.

The terminology and definitions used in the ECSC are similar to, but not always identical with, those in the ECC

Some of the more important terms are described in the following. Reference should be made to the chapter on the ECC for further details of the definitions.

Completion It is possible for the works to be completed as defined but containing defects. The test as to whether defects in the 'completed' works are acceptable is whether or not those defects will prevent the Employer from using the works, or other parties from doing their work. Work still to be carried out in a part of the works (outstanding work) may be regarded as a defect in that part of the works. The definition of completion largely removes any subjective element in judging whether or not the works have been completed.

Under clause 30.2, the Contractor is required to submit to the Employer a forecast of the date of completion on a weekly basis. In the absence of a programme, or where periodic revisions of the programme are not required, the estimated date of completion is very useful to the Employer.

Defined Cost This term has been changed from the 'Actual Cost' of the previous edition. It is used only for the assessment of compensation events, which include items of work for which there is no rate in the Price List. The four components listed replace those in the schedules of cost components of the ECC. This simplification is made possible because of the absence of target and other cost reimbursement options in the ECSC. Often the most controversial component is that resulting from the pricing of contractor's equipment. This is dealt with by using market hire rates.

Equipment, plant and materials The definition of Equipment covers anything of a temporary nature. Plant and materials are those things incorporated in the final works and that are not removed at the end of the contract.

Price for work done to date This term determines the interim and final amounts to be paid to the Contractor. It consists of two parts – lump sums and remeasured items from the Price List. Throughout the conditions there are provisions for other amounts to be paid to or by the Contractor and these also must be taken into account in calculating amounts due.

Prices These are the amounts in the final column of the Price List inserted by the Contractor at tender, and which form the basis of the financial part of the contract. The financial consequences of compensation events are assessed in terms of the impact of the events on the Contractor's costs (Defined Cost).

The ECSC contains provisions that are similar to but not identical with those in the ECC. Some of these are described in the following

Communications The detailed provisions of clause 13 of the ECC are much less likely to be required in a short contract. Hence, only two have been included, namely the need for all valid communications to be in writing, and the requirement that replies should be within the times stated in the contract. The sanction available to the Contractor if the Employer fails to reply within these stated times, is notification of a compensation event.

Early warning This important procedure is an abbreviated form of that of the ECC. It omits the details of the risk reduction meeting with its specified agenda, but includes the basic requirement for both parties to co-operate in seeking a solution to the problem that has been notified.

Subcontracting The Contractor is not required to seek acceptance of the names and conditions of contract in respect of work that he intends to subcontract. However, clause 21 makes clear that the Contractor is responsible, whether or not work is subcontracted. In the absence of the ECC's target and cost reimbursement options, the identity of subcontractors and their terms of subcontract assume lesser importance.

Managing defects The procedure for dealing with defects begins with the Employer notifying the Contractor of them. The Contractor has an overriding obligation to correct defects whether or not they are notified in this way. Timing of the correction depends on whether or not failure to correct the defect will prevent the Employer or others from doing their work. But, where defects are notified only after completion, they are to be corrected within a stated time period, namely the defect correction period stated in the contract data. There is no mechanism stated in the contract to enable the Employer to accept a defect, as in the ECC. However, the parties are always free to negotiate and agree terms under which a particular defect does not have to be corrected. This is implied as an exception in the first compensation event listed under clause 60.1

Payment Procedures and timing of payments to the Contractor are essentially different from those of the ECC, reflecting the nature of the short contract. In the ECSC, timing of payments is related to an 'assessment day', which is stated in the contract data as being a particular date in each calendar month. The Contractor is required to submit to the Employer an application for payment by each assessment day. This contrasts with the ECC under which the submission of an application by the Contractor is not obligatory, and where the onus of assessing the amount due is on the Project Manager. Where the Employer disagrees with the amount applied for, he makes the necessary corrections and advises the Contractor accordingly. Where the Contractor fails to complete his work by the completion date, delay damages at the rate stated in the contract data are deducted in calculating the amount due. Retention is also deducted as a percentage of the price for work done to date. Payment is to be made within three weeks of the assessment day – there is no express provision for other periods. However this period could be amended by an additional condition in the contract data.

Compensation events In contrast to the 19 compensation events listed in the core clauses of the ECC, there are only 14 listed in the ECSC. None of these refers to dates shown in the programme, suggesting that dates in the programme (if any) have a lesser contractual status. However, there are other restraints on the Employer albeit expressed in different terms. Many of the compensation events represent breaches by the Employer of his contractual obligations. The risk of unforeseeable physical conditions has again been placed on the Employer as in the ECC. The basis of judging such conditions is the same as that stated in the ECC, namely on prior visual inspection of the site by the Contractor and other factors. The weather risk is shared between the Contractor and Employer. However the threshold up to which the Contractor takes the weather risk is stated in rather different terms from those of the ECC. The ECC describes the threshold on a statistical basis using 'weather measurements'. By way of contrast, the ECSC threshold is established using the number of full days during which all work on the site is prevented from being carried out. This is expressed as a proportion of the total number of days in the contract period. It seems likely that this threshold is rather more onerous on the Contractor than in the case of the ECC and, if this is correct, it means that the Contractor carries a greater risk of bad weather under the ECSC.

The penultimate compensation event is described as a difference between the final total of work done and the quantity of an item stated in the price list. At first glance this seems to be unnecessary, bearing in mind that quantities in the price list are remeasured, and compensation events are assessed using the rates in the price list. However, all compensation events are assessed not only on the basis of the financial effect but also on the effect on planned completion resulting in possible delay to the contractual completion date. Thus, where the quantity of an item has increased, and that item of work is critical to completion of the works, a delay to the completion date (commonly called extension of time) may be due to the Contractor.

The procedures for managing compensation events are described in detail, as in the similar provisions of the ECC. This probably reflects the importance of managing change effectively, even in a short contract. It is likely that most of the administration load of a contract under the ECSC relates to compensation events. Both parties may notify the other of the occurrence of a compensation event although notification by the Employer arising from an Employer's instruction is only implied. Failure by the Contractor to notify a compensation event within eight weeks of becoming aware of it invokes a time-barring sanction of loss of entitlement. The financial part of assessing compensation events is rather different from that in the ECC. Where there are items and rates in the Price List for the work in question, those rates are used for assessing the event. Where there are no such rates, compensation events are assessed on the basis of effect on cost – either historical cost, where the event has already occurred, or forecast cost, where the effects of the event have yet to occur. Assessments of compensation events are final except in the case of qualified quotations based on assumptions as instructed by the Employer. Thus, once the assessment is fixed, the Contractor takes the financial risk from that time onwards. In consequence, the certainty of the amount of the payment is of benefit to both parties.

Disputes which the parties themselves cannot resolve may be referred by either party to the Adjudicator

Adjudication, as a first means of resolving disputes has become well established in the UK construction industry, largely as a result of the Housing Grants, Construction and Regeneration Act 1996 (the Act). The procedure for referring disputes under the ECSC to the Adjudicator is described in section 9 of the conditions of contract. The procedure is inquisitorial in that it entitles the Adjudicator to take the initiative in ascertaining the facts and the law. The Adjudicator is required to give his decision together with reasons within four weeks or a longer agreed period. A dissatisfied party may refer the dispute afterwards to the 'tribunal', which is normally arbitration or litigation in the courts. Strict time limits are stated for referring the dispute to the arbitrator or the tribunal. The Act, however, enables a party to refer a dispute to adjudication at any time. Thus, if the Act applies to the work in question, the ECSC includes a replacement clause to make the procedure comply with the Act. For this reason, it is necessary to state against the appropriate entry in the contract data whether or not the Act applies. Examination of the Act will make clear whether or not the Act applies.

When to use the ECSC

From the above, the choice as to whether to use the ECSC or the ECC may be straightforward, but in borderline cases the choice may be a difficult one. Previous experience in the use of both contracts will make the choice easier. Much will depend on such things as the complexity of the work, the risks involved and the simplicity or otherwise of the management procedures necessary to administer the contract. Where the risks are substantial and in the time available cannot be reduced, a target or cost reimbursement contract may be the most appropriate form. If this is the case, the ECC should be used since the ECSC does not cater for this type of contract.

How to use the ECSC

In preparing the tender documents, decision making is much more straightforward than in the case of the ECC. However, much care and effort is required to prepare the Works Information properly, as the success and smooth running of the project may depend on how well this task has been carried out. The procedures for inviting tenders and creating a contract with the successful contractor is fully described in the published guidance notes. Once the contractor is appointed it may be prudent to run training sessions for all personnel who are to be working on the project. This will assist in familiarising the people belonging to both parties with the content and procedures of the contract (the ECSC may be new to many of them). It will also start the process of team building and encouraging the adoption of right attitudes resulting in mutual trust and co-operation – a vital requirement of an NEC contract. Hard work may be required by both parties to change the frequently entrenched attitudes engendered by more traditional forms of contract.

Exercises

(1) List the main risks that you would expect to find in the works listed below, and then decide whether you consider that the short contract (ECSC) would be suitable. The estimated cost is shown in brackets in each case. Make any necessary assumptions.

  (a) Demolition of a nine-storey hotel building with a reinforced concrete frame in an urban area, and construction of a temporary car park for the local authority on the levelled site. [£250 000]

  (b) Widening an 800 m length of dual carriageway and constructing a bus lane. [£620 000]

  (c) The sewerage system serving an old housing estate is in a bad state of repair. It has been decided to construct new sewers and manholes adjacent to the existing sewers. [£680 000]

  (d) Demolition of a prestressed concrete footbridge over a motorway and construction of a replacement footbridge in structural steel. [£950 000]

  (e) Supply and installation of 10 km of high-pressure steel gas pipeline 1.1 m dia in a trench excavated by another contractor. [£2 750 000]

  (f) A six-bedroom house has suffered extensive cracking due to settlement caused by surrounding poplar trees and other vegetation. All the load-bearing walls are to be underpinned to depths of between 1 and 2 metres. [£75 000]

  (g) A sports club wishes to build a new spectator stand with seating accommodation and refreshment facilities for its athletics and football activities. [£420 000]

  (h) A canal authority proposes to replace the lock gates, pumps and sluice valves, and repair the brick walls at three locks on its canal system. The canal is used only for leisure purposes. [£800 000]

  (i) A local authority wishes to build a new paved area for a permanent car park on the site of an old gas works which was demolished to ground level some ten years previously. [£140 000]

  (j) A highway authority proposes to install a new roundabout at a junction of five roads in an urban area. Two sets of pedestrian traffic lights are also required. [£680 000]

  (k) A housing trust wishes to refurbish the kitchens and bathrooms in 500 residential units that it owns. The replacement units and furniture are to a standard design. [£8 m]

5 The Engineering and Construction Short Subcontract (ECSS)

Origins

The first edition of the ECSC was published in 1999 together with guidance notes and flowcharts. Those guidance notes included a section entitled 'Use of the ECSC as a subcontract'. This described how the ECSC could be amended for use as a short subcontract. Thus, a contractor wishing to use it as a subcontract would have needed to physically amend a copy of the ECSC, or issue a list of amendments to be read with the ECSC. Neither of these was likely to produce a user-friendly document for subcontract work. It was probably for this reason that a first edition of the short subcontract (ECSS) was published in 2001. When the family of NEC3 documents was published in 2005, second editions of both the ECSC and ECSS were published, together with a set of guidance notes covering both contracts. At the end of these guidance notes there was a separate short section entitled 'Supplementary guidance notes for the ECSS'. This recognised the similarity between the two contracts.

The ECSS is not restricted to use as a subcontract to the ECSC

The ECSS is most likely to be used where work under the short contract (the ECSC) is subcontracted to a subcontractor. As a result of the general back-to-back nature of the two contracts, this ensures that the obligations of the Contractor under the ECSC are largely passed to the Subcontractor, establishing a chain of liability. But the ECSS may be used in other situations. Its most common alternative use is likely to be for subcontract work under the ECC. Many subcontracts under the ECC are for straightforward work, involving low risks and unsophisticated management procedures – characteristics that the ECSC and ECSS were designed for. These smaller contracts are unlikely to justify use of the Engineering and Construction Subcontract (ECS). Many of the contract conditions of the ECC are reproduced in the ECSC and ECSS, albeit in an abbreviated form. Thus, while the back-to-back nature of these two contracts is less evident, there should be little conflict between the two. In such a situation the subcontract data of the ECSS may require rather more careful drafting.

One of the newer contract forms published under NEC3 is the Term Service Contract (TSC). This is essentially different from others in the NEC family in that it caters for providing a service over a fixed period of time, rather than providing a project by a completion date. It is designed for use for maintenance-type work. However, in carrying out such work the contractor may need to carry out discrete items of work as part of providing the service. Examples are replacement of things that have reached the end of their useful lives, and providing project-type works under the Task Order option of that contract. Except for major high-risk work carried out under the TSC, much work that the contractor wishes to subcontract is of such a nature that the ECSS is ideally suited for that purpose.

A further, though less frequent, use of the ECSS is as a subcontract under the Professional Services Contract (PSC). Sometimes, when a consultant is engaged to do a feasibility study or design work, some physical investigation work is necessary. For example, this may comprise the provision of physical access to examine the condition of a bridge or building, or a ground investigation with boreholes, trial pits and soil testing. The consultant will normally arrange for this type of work to be done by a contractor and the ECSS is well suited for this purpose. The consultant will fulfil the role of the contractor and some minor amendments to the contract data of the ECSS will be necessary.

Completion of the forms in the ECSS is very similar to completion of the forms in the ECSC

There is a certain amount of additional information included in the contract data of the ECSS, mainly concerning the main contract details. For example, the names and details of the Employer, Project Manager, Supervisor and Adjudicator in the main contract are to be stated where applicable. Also included is a description of the main contract works as well as the subcontract works. Where the main contract is the ECS, TSC or PSC, these entries will need to be amended to suit. There are obvious advantages in having the same person named as adjudicator in both the ECSC and ECSS.

The period for reply is related to the period for reply in the main contract. Replies by the Contractor to the Subcontractor are to be a stated number of weeks more than the period for reply stated in the main contract, to allow for the passage of information emanating from the Employer. Similarly, replies by the Subcontractor to the Contractor are to be a stated number of weeks less than the period in the main contract. This allows the Contractor time to pass the information to the Employer and so fulfil his obligations for timing of his reply under the main contract.

Careful consideration is needed in deciding the defects date of the subcontract. There are two possibilities. First, it may be a stated period after completion of the subcontract works. Thus, if the subcontract works are completed some time before completion of the main works, the Contractor carries certain risks relating to the subcontract works from the defects date of the subcontract to the defects date of the main contract. Second, it may be related to completion of the main contract, in which case the Subcontractor carries risks of the completed subcontract works for an extended period.

Pricing of the Price List is entirely a matter between the Contractor and Subcontractor and does not relate directly to the Price List in the main contract. However, pricing of compensation events in the main contract may involve rates and prices in the subcontract price list where assessment has to be based on the effect of a compensation event on defined cost (the latter includes effects on defined cost of subcontracted work).

The Works Information is a detailed statement of what is to be provided by the Subcontractor and how, and what things and facilities, if any, are to be provided by the Contractor

At first glance the Works Information section appears to be identical to that of the ECSC and, in preparing tender documents for a subcontract, the Contractor may well be tempted to copy all the main contract information. This may result in much irrelevant information being included, leaving it to the Subcontractor tenderers to identify what information applies to the sub-contract works. If problems are to be avoided, the Works Information should be carefully drafted. While much of the information may be reproduced from the main contract documentation, there may be additional information required. In particular, the relationship of the subcontract works to the main contract works should be made clear, and how the timing of the subcontract work fits into the main contract programme. One of the entries under the section on programme requirements concerns the use for which the sub-contract works are intended. This statement is for use mainly to decide when the subcontract works have been completed. Section 6 should list all the things that the Contractor intends to supply for the Subcontractor's use, such as services, equipment including health and safety equipment, and welfare facilities.

There are no named parties in the ECSS conditions other than the Contractor, Subcontractor and Adjudicator

The main powers and obligations in the ECSS are exercised by the contracting parties themselves – there is no equivalent of the Project Manager or Supervisor. Neither is there any provision for delegation as in the ECSC. However, at the start of the subcontract, it is good practice for the parties to state who will be representing them and what powers they are given.

The terminology, definitions and provisions in the ECSS are similar to, but not always identical with those in the ECSC

Some of the more important terms and provisions are described in the following. Reference should be made to the chapters on the ECC and ECSC for further details of the definitions. Certain terms have been prefixed with 'subcontract' to distinguish them from terms in the ECSC, and a subcontractor of the subcontractor is described as a subsubcontractor.

The weather compensation event is similar to that in the ECSC but it may have different consequences. Many subcontracts will be of shorter duration than that of the main contract. The occurrence of a weather compensation event is expressed, as in the ECSC, in terms of the number of full working days lost to bad weather as a proportion of the total number of days in the contract. Hence, the ECSS appears to be more sensitive to the occurrence of a weather compensation event than the ECSC because of the possible shorter period of the subcontract. It is possible that a weather compensation event may occur in the subcontract but not in the main contract. The Contractor should be aware of this small difference of risk.

A further compensation event has been added in the ECSS. This concerns the possible extended period from completion of the subcontract works to the defects date, where the latter is the same as the main contract defects date. The inclusion of this compensation event has the effect of removing from the Subcontractor the risk of the consequences of late completion by the main contractor.

The ECSC and ECSS both contain a mechanism whereby a particular dispute affecting both contracts may be referred to a single adjudicator. He then decides the dispute as between all three parties. To make this effective the recommended additional clause in the ECSC guidance notes should be incorporated as an additional clause, The reciprocal clause is already included in the ECSC under the heading 'combining procedures'.

When to use the ECSS

From the above, use of the ECSS as a subcontract in the ECSC is an obvious choice, because of its simplicity and back-to-back nature. Use of the ECSS as a subcontract in the PSC has many advantages where work of a physical and manual nature is required. Similarly, use of the ECSS as a subcontract in the TSC has distinct advantages over, say, the ECSC, since the former is drafted as a subcontract rather than a main contract.

How to use the ECSS

Notwithstanding the fact that this contract is largely back to back with the ECSC, great care should be exercised in its preparation. Once a subcontractor is appointed, it may be prudent to invite his personnel to join in training sessions for all people working on the project. This should assist in familiarising all parties with the content and procedures of the contracts and developing team working. The obligation of mutual trust and co-operation is a basic requirement of the ECSS as well as the other NEC contracts; joint training sessions should help all parties to meet these obligations.

Exercises

(1) The local authority of a coastal town proposes to invest in a leisure development scheme to attract tourists. It has placed a management contract valued at £35 m with a contractor under the ECC using main option F. The various works packages are to be let as subcontracts. Decide which of the following items of work would be suitable for use with the short form of subcontract (ECSS) giving reasons for your answer. The estimated value of each package is shown in brackets. Make any necessary assumptions.

(a) Dredging of a disused harbour to provide 3 m minimum depth of water for mooring yachts. The dredging may include some underwater rock excavation. [£450 000]

(b) Site clearance consisting mainly of demolition of derelict buildings and structures. [£550 000]

(c) Drainage, road works and service ducts to provide the infrastructure for the development and access from the existing roads. [£4.5 m]

(d) Design and construction of a yacht club building. [£2.5 m]

(e) Design and construction of 15 residential apartments. [£4.5 m]

(f) Design and construction of a children's paddling pool. [£1.2 m]

(g) Design and construction of six kiosks located on the promenade. [£35 000 each]

(h) Landscaping. [£400 000]

6 The Professional Services Contract (PSC)

Origins

The first edition of the PSC was published in 1994, which was one year after publication of the first edition of the NEC, as the 'black book' was then called. The original demand doubtless arose because of the need to appoint a Project Manager and Supervisor under the NEC. Thus, the PSC was drafted to cater for Employers who wished to appoint professionals, using the same principles as were used in drafting the NEC. Most standard forms of contract for professional appointments have hitherto been written by major employers, trade organisations or professional bodies but rarely by bodies representing all sides of the construction industry. It is evident that people from all sides of the industry took part in the drafting of the PSC, presumably with the object of achieving 'fairness' and an appropriate allocation of risk between the parties. Although disputes in professional work have generally occurred less frequently than in construction work, nevertheless it seems likely that the NEC approach will reduce them even further.

The publication of the PSC in 1994 had already anticipated the recommendation of Sir Michael Latham in his report 'Constructing the Team' for a wholly interrelated package of documents.

The second edition of the PSC was published in 1998. This incorporated most of the changes introduced into the post-Latham (and second) edition of the NEC, then renamed the Engineering and Construction Contract (ECC) and published in 1995. Its scope was also expanded to cater for appointment of professionals of any kind – both within and outwith the construction industry. As the industry gradually became more familiar with the NEC system, use of the PSC increased in both public and private sectors. The third edition of the PSC was published in 2005 under the general title NEC3. As in the case of the other NEC3 standard forms of contract, the PSC has received the official endorsement of the UK's Office of Government Commerce (OGC) in that it fully complies with the *Achieving Excellence in Construction* (AEC) principles. Accordingly, the OGC recommends its use by public sector procurers in construction projects. Publication of revised editions of all the NEC contracts at the same time, allowed for a greater degree of consistency and hopefully facilitates familiarity with the various conditions of contract.

The essential and distinguishing features of the PSC

The published document As in the case of other NEC forms of contract, the published PSC document is not a contract but rather a number of statements and provisions from which a contract to suit particular circumstances may be prepared. An informed Employer should have little difficulty in preparing a contract under the PSC for the appointment of a Consultant, in spite of the fact that much thought and effort may be required to achieve a satisfactory result. A less experienced Employer may need help to do this.

A contract under the PSC will comprise:

(a) Core clauses. These must be included in every contract.
(b) One main option selected from the four main options: A, C, E and G. Selection of the main option determines the allocation of risk between the parties and how the Consultant is paid. The numbering of the

clauses follows on from the core clauses and under the same side headings as in the core clauses.

(c) A dispute resolution option W1 or W2, according as to whether the UK's Housing Grants, Construction and Regeneration Act 1996 applies to the contract.

(d) Secondary options selected from the numbered options X1 to X13, X18 and X20, and the two options numbered Y(UK)2 and Y(UK)3. This selection determines the allocation of further risks.

(e) Additional bespoke conditions of contract indicated by the letter z. These are normally decided by the Employer or by agreement between the parties. Because of the options available, the need for additional conditions should be minimal.

(f) Contract Data. This is in two parts, part 1 being prepared by the Employer and part 2 prepared in part by the Employer for completion by tendering consultants or a selected consultant. The contract data contains information specific to the contract. Some information in the contract data takes the form of reference to other documents, which are thereby incorporated into the contract.

One of the most important decisions the Employer has to make is to decide which main option to use

There are four main options: A, C, E and G. The Consultant carries the greatest financial risk under option A and the least under option E. Selection is determined by the nature of the work that the Consultant must carry out, and how the risks are to be dealt with. Conspicuous by its absence is an *ad valorem* option under which a consultant is paid for design work calculated as a certain percentage of the works' construction cost. This has been the basis of payment under many consultancy agreements for many years. The guidance notes list a number of reasons why this option has not been included.

In each option, the Consultant is paid expenses as defined in the contract, in addition to the payment for services provided, calculated in accordance with the main option.

The lump sum option. Option A: Priced contract with Activity Schedule

This is very similar to option A of the ECC. It is a lump sum contract in which the Consultant undertakes to provide the services described in the contract for a sum of money. He carries the risk of providing those services for the lump sum. The Activity Schedule is a list of activities to be carried out by the Consultant in providing the services. This is normally written by the Consultant since he is the one who knows what activities will be carried out, but the Employer may specify the form it should take. Each activity is priced by the Consultant as a lump sum, which is the amount paid to the Consultant when he has completed the activity. Because of this payment condition, consultants should ensure that the number and description of items in the schedule are such as will maintain an acceptable cash-flow. The total of the lump sum prices is the Consultant's price for completing the whole of the services. In pricing an activity, the Consultant takes responsibility for estimating and pricing the resources he is likely to need making due allowance for risk, overheads and profit.

The lump sum basis of payment has not been commonly used in the past as a basis for paying for consultancy services. However, provided that the scope of the work is well defined and the Consultant is sufficiently experienced, there is no reason why the Consultant should not be able to estimate and price the required resources needed with reasonable certainty. It has the advantage for both parties of being easy to administer.

The target option with the target established as a lump sum. Option C: Target contract

Under this option, the Consultant tenders or negotiates a target price (defined as the Prices) using an activity schedule. Unlike option C in the ECC, there is no fee included for use in calculating payments to the Consultant. This means that the staff rates tendered in part 2 of the contract data must include for the Consultant's overheads and profit. During the course of the contract, the

Consultant is paid a time charge, which is the number of hours spent by the Consultant's staff, priced at the rates in the contract data. At the end of the contract, the final time charge is compared with the target. If this shows a saving, the difference is shared between the Employer and Consultant, in proportions which are defined in the contract. If it shows that the target price has been exceeded, the Consultant is required to pay back a proportion of the excess. The Consultant's share is calculated on two occasions – at completion, when the calculation is based on a forecast of the final figures (for target price and time charge), and at the final account stage, when the calculation is refined using the final figures. During the course of the contract, the target price is adjusted to cater for compensation events – these are events that are defined in the contract as being at the Employer's financial risk.

Thus, option C is basically a time-based contract, which incorporates an incentive for the Consultant to minimise costs. Savings and overruns are shared between the parties. In risk terms it lies between option A and option E. It is appropriate to use option C in cases where a lump sum contract would place too much risk on the Consultant but where a purely time-based contract would place too much risk on the Employer, and thus provide little incentive for the Consultant to work efficiently. For this reason the appropriate use of option C is likely to reduce the occurrence of problems and disputes.

Hitherto, there have been few standard conditions of contract for target contracts, but the increasing use of them for construction work may result in their increased use in contracts for the provision of professional services. It may be suitable for contracts which previously have been carried out on an *ad valorem* or fee percentage basis.

The option where the Consultant is paid for the time spent. Option E: Time based contract

Under this option the Consultant takes very little financial risk. It is used where the work to be done by the Consultant cannot easily be defined at the outset and when the financial risks of doing the work are great. It is therefore suitable for engaging a Consultant to carry out a feasibility study, or to supervise the construction of work being carried out by a contractor. In the latter case, if the construction is being carried out under the ECC, the PSC would be for the appointment of the Consultant as Project Manager and Supervisor. This option may also be suitable for such varied tasks as studies of alternative designs, investigation and reporting on structural failures, research and experimental work, work as an expert witness, advising on the condition of existing buildings structures or sites, advising on procurement and contract strategy, and developing alternative designs for a contractor under a design and build contract. Control of final cost is maintained by requiring the Consultant to submit to the Employer forecasts of the final total time charge and expenses, periodically throughout the contract.

The option where an Employer wishes to retain the services of a Consultant on a call-off basis over a period of time. Option G: Term contract

Sometimes an Employer wishes to retain the services of a Consultant to advise on specific matters or carry out particular tasks. Many of these tasks may be relatively minor in extent, such as would not justify agreeing a separate contract in each case. Thus, this option may be regarded as a standing offer by the Consultant to carry out work of a certain kind at the prices agreed in the contract. The Employer instructs each task in the form of a task order, which states details of the task, timing and pricing. Pricing of each task is based on prices in a task schedule. This schedule is a list of items of work likely to be required by the Employer, each of which is priced as either a lump sum or stated to be time based. Items of work that are not on the schedule, are priced as compensation events, that is as a forecast of the time charge.

Interim payments are calculated using the number of hours spent on the time-based items and an appropriate proportion of the prices for the lump sum items.

Control of costs is exercised by periodic forecasts of final costs as in the case of the time-based option. Clearly, forecasting is more difficult in this case when the future requirements of the Employer are not known.

The dispute resolution options determine the procedures for dealing with disputes

The dispute resolution options W1 and W2

As in other NEC contracts, the parties are required to act in a spirit of mutual trust and co-operation. This fiduciary duty has always been a characteristic of professional work, but in the PSC it is expressly stated in the first clause of the conditions. Nevertheless, disputes do arise for various reasons and it has been the practice for many years to include in conditions of contract procedures for resolving disputes without recourse to the courts. The method of dispute resolution, specified in the past, has usually been arbitration. However, the NEC introduced adjudication as the initial method of dealing with disputes, and in the UK this has since been made a statutory right by virtue of the Housing Grants, Construction and Regeneration Act 1996 (the Act).

The NEC3 has introduced two dispute resolution options:

- Option W1 for contracts not subject to the Act.
- Option W2 for contracts that are subject to the Act.

The main provisions of these options have been described in the chapter on the ECC.

Since the Act applies only to construction contracts, it may appear that it has no application to contracts for providing professional services. But this is not so.

As defined in the Act, construction contracts relate, among other things, to the carrying out of construction operations or arranging for the carrying out of construction operations by others. It includes an agreement to do architectural, design, or surveying work, or to provide advice on building, engineering, interior or exterior decoration or on laying out of landscape, in relation to construction operations. The Act includes a long definition of construction operations as well as a definition of certain activities that are not construction operations. It is clear, from the definitions in the Act, that many consultancy agreements will be subject to the Act but, where there is doubt in any particular case, the Act should be consulted.

The adjudicator is named in the contract data and should be a person with some knowledge and experience of consultancy work.

Various other risks need to be considered to decide how they can be reduced and which party should carry them so that the likelihood of disputes arising is minimised. These risks are catered for in the secondary options

Which party should carry the risk of price increases? Option X1: Price adjustment for inflation

The Employer must make the decision on which party is to take the risk of salary rises of the Consultant's staff. For contracts of short duration or in countries where inflation is relatively low, it should not be too difficult for the Consultant to allow, in his staff rates, for future increases of salary. But in other circumstances, it may be preferable for the Employer to take the inflation risk by including this option in the contract.

Since the main element of a Consultant's costs is that of staff salaries, this option provides for two situations. The first of these is where the staff rates stated in the contract data are fixed in the contract and do not change even if the actual salaries paid to individuals increase. The second is where the staff rates stated in the contract data are variable in line with the changes in salary paid to individual staff members. The price adjustment factor is calculated on a similar basis to that in the case of the ECC, using indices appropriate to the PSC. It is calculated after each complete year of the contract and is applied throughout the year following. Different clauses apply depending on whether staff rates in the contract data are fixed or variable. Compensation events assessments for options A, C and G are adjusted back to base date levels before the formula is applied. Expenses also are adjusted using a price adjustment factor, unless they automatically take account of price rises.

For comments on options X2 to X7 the reader is referred to the comments on these options in the chapter on the ECC.

By agreement with the Consultant, the Employer can pass some of the benefits of the contract to third parties. Option X8: Collateral warranty agreements

The system of collateral warranty agreements is the mechanism by which a contract is created between the Consultant and third parties whom the Employer may wish to benefit directly from the contract. The commonest situation where this applies is where the Employer is a developer who wishes to create legal liabilities between the Consultant and, say, funding organisations or purchasers or tenants. Collateral warranties create additional risks and liabilities on the part of the Consultant and this, in turn, should be reflected in the terms of payments to the Consultant. The precise terms of collateral warranties are clearly of considerable importance to all parties. There are a number of standard forms of warranty in existence and these contain varying degrees of risk for the Consultant. Those published by the Construction Industry Council and the British Property Federation are widely used.

For work in the UK an alternative to this option is option Y(UK)2, by which third parties may acquire rights in the contract as a result of the Contracts (Rights of Third Parties) Act 1999.

The Employer may acquire from the Consultant rights (intellectual property rights) over material produced by the Consultant. Option X9: Transfer of rights

Normally the rights over material produced by a party remain with that party. The core clauses in section 7 recognise this but they do also provide for the Employer to have certain limited rights of use (licence to use). In the same way, these clauses also restrict the right of the Consultant to use material provided by the Employer. However, the effect of including option X9 is to transfer the rights of the Consultant over material he produces to the Employer but subject to any provisions in the Scope.

This is an important addition to any contract for professional services, particularly where the Employer may wish to use a Consultant's design for further work or for extensions and alterations in the future.

For practical reasons, an Employer may not be able to carry out his duties under the contract, and may wish to delegate powers to another. Option X10: Employer's Agent

There is no provision, in the core clauses of the PSC similar to that in the ECC, for the Consultant to delegate his authority under the contract. This option provides the means to enable the Employer to delegate his authority to an individual or to another consultant to act on his behalf.

There are a number of reasons why an Employer may wish to terminate a contract. Option X11: Termination by the Employer

Strictly, this option is not concerned with ending the contract, since the contract itself continues in existence for some time. Its use is limited to terminating the Consultant's main obligation, which is to provide the services. The main provisions entitling a party to terminate are in the core clauses of section 9. Without these clauses, termination by one of the parties would be a breach of contract. But the PSC, in common with most standard contracts, provides the parties with entitlement to terminate in certain circumstances. The procedures and payment following termination depend on the reasons for termination. The reasons stated in the core clauses are fairly wide-ranging and include, for example, a situation where the Employer no longer requires the services. Inclusion of option X11 widens even further this right of the Employer, namely to terminate at will. But there is a price to pay for this right to 'walk away' and that is a payment additional to that described in the core clauses. It seems to recognise loss of profit arising from the termination. It is possible that the right in this option may be exercised by an Employer who still requires the services but for some reason wants them to be provided by another consultant.

How can the Consultant be encouraged to work with other parties on the project, as a team to produce a successful outcome? Option X12: Partnering

This option is similar to that in the ECC and its provisions are discussed in the chapter on the ECC. On any particular project there may be several consultants, all of whom are partners. Not all, however, may be included in the core group of partners. Where one of the consultants acts as a lead consultant, it is likely that he would be a key member of the core group together with others who fulfil an important role in the project.

For comments on option X13 (Performance bond) the reader is referred to the chapter on the ECC.

Should there be some financial limit to the Consultant's legal liability? Option X18: Limitation of liability

The nature and extent of the services to be provided by a Consultant may vary considerably. In some cases the liability he incurs may be considerable and that is why the core clauses (section 8) provide for some limitation to the total liability. There are some significant exclusions listed in the core clause. The financial limit stated in the contract would normally be the same as the insurance cover. Further limitations to the Consultant's liability are stated in option X18. These relate to the Employer's indirect or consequential losses and what is commonly called latent defects. This option recognises the difficulties a Consultant may experience in obtaining insurance cover on commercially reasonable terms. The option also includes a time limit to the Consultant's liability similar to that in the ECC.

Should the Consultant be rewarded for meeting specified targets? Option X20: Key Performance Indicators

An important aspect of recent commercial and industrial life has been the introduction of targets and incentives with the object of achieving continuous improvement. Incentives are now commonly used, not only to reward contractors but also consultants in their provision of professional services in order to encourage better efficiency of performance. The purpose of the Key Performance Indicators (KPIs) is to provide such an incentive. Performance is measured in different ways and the Consultant's actual performance is compared with targets to determine what financial rewards are due to him.

What is necessary to make the contract comply with UK legal requirements? Option Y(UK)2: The Housing Grants, Construction and Regeneration Act 1996

The construction part of this Act was introduced to deal with two matters, namely adjudication of disputes and payment. The first of these is dealt with by selecting the appropriate Dispute Resolution Option W2 and payment is dealt with by incorporating this option Y(UK)2. It should be used only where the Act applies to the work in question. How to decide whether or not the Act applies, is discussed above, in option W2. This option introduces two new terms as required by the Act, namely a 'due date' and a 'final date' for payment. Payment under the PSC is initiated by the Consultant's invoice and the due date is seven days after the date of this invoice. The final date for payment is 14 days after the due date (or other period stated in the contract data). Having received the Consultant's invoice, the Employer is required to give notice to the Consultant of the amount he intends to pay. This notice is to be given within five days of the date on which payment has become due. If for some reason the Employer intends to withhold payment of an amount due, he must notify the Consultant. Failure to do so entitles the Consultant to suspend performance. If he exercises this right, he is entitled under the Act to an extension of time equal to the period during which payment is withheld. Additionally under Y(UK)2 he is entitled to additional costs, as his proper exercising of his right to suspend performance constitutes a compensation event. The purpose of the Act is to provide a practical remedy for the Consultant who is not paid the amount due at the appropriate time. Prolonged delay in paying the Consultant entitles the latter to termination.

Should third parties have any rights even though they are not parties to the contract? Option Y(UK)3: The Contracts (Rights of Third Parties) Act 1999

This Act permits third parties to exercise rights under a contract even though they are not parties to that contract. Any such specific rights that the parties agree to provide for are to be stated in the contract data.

Is any fine tuning required to meet the requirements of the Employer and the Consultant and to suit the circumstances of the project? Option Z: Additional conditions of contract

In spite of the flexibility available in the structure of the PSC, it will often be necessary to make additional provisions to suit the circumstances of a particular situation – the fine-tuning. Such additional z clauses comprise the bespoke element of the contract and are usually decided by the Employer but sometimes negotiated between the Employer and Consultant. Additional clauses should not be used to make substantial changes to the overall balance of risk between the parties as this may cause conflict within the contract. Additional clauses should be carefully drafted to harmonise with the standard clauses, avoiding overlap wherever possible.

Roles, duties and powers of the Employer and the Consultant are described in detail throughout the conditions of contract

The contracting parties are the Employer and the Consultant, each of which has specific powers and duties described in the contract. If secondary option X10 is included in the contract, the Employer may delegate his authority to an agent. The only other named party identified in the contract data is the adjudicator. His function is to decide disputes referred to him by one of the contracting parties, under option W1 or W2. Subconsultants are defined in the contract, but not named. 'Others' is also a defined term. The roles, duties and powers of these various parties are described throughout the conditions but section 2 covers the main responsibilities of the Employer and Consultant.

The basic duty of the Employer and Consultant is stated in the first clause of the contract. This describes not only what these parties must do but also how they are to act. The first of these is inherent in the nature of contracts. The various actions are generally described in the present tense and the use of the word 'shall', in the first clause, makes the actions mandatory. The second duty expressly states that the parties are to act in a spirit of mutual trust and co-operation; this was included following a recommendation in the Latham Report. The element of trust is basic in appointing a Consultant to provide professional services. It has always been one of the main factors that distinguishes a professional service from a non-professional service. Hence, compliance with this first clause should not prove to be onerous.

The Consultant and Employer have many fundamental obligations

The Consultant's main obligation is to provide the services, and the detail of what and how he is to provide it is described in the scope. The standard of care that he is required to exercise in providing the services is that of using the skill and care normally used by professionals providing services of a similar kind. It is common practice in contracts for professional services to specify this standard – usually referred to as reasonable skill and care or reasonable skill, care and diligence. This contrasts with the higher level of care described as fitness for purpose.

An Employer may be much more interested in the identity of the individual people working on his contract, than in the Consultant and his reputation as a firm. It is for this reason that allowance is made for the Consultant's key people to be named in the contract. Whenever individual names are included in a contract, however, there must be provision for their replacement if this becomes necessary. The criteria on the basis of which the Employer may reject a proposed replacement are stated, namely that the replacement person must have equivalent qualifications and experience.

The Consultant has a duty of co-operation, not only in relation to the Employer but also in obtaining from, and providing information to, others. The Employer is entitled to attend any meetings with others, including other consultants engaged in the same project, arranged by the Consultant, thus ensuring considerable transparency in relationships. Where partnering is a formal part of the contract, under option X12, communication with other partners is detailed in more specific terms in the Partnering Information.

In this third edition of the PSC, the concept of key dates has been included as in the case of the ECC. This places a specific obligation on the Consultant to meet a stated condition of a part of the services by a particular date, called a key date. The sanction for failure by the Consultant to meet a key date is stated as being the additional cost incurred by the Employer.

The Consultant may subcontract some of the services but this is subject to the Employer's acceptance. Where certain work to be carried out by the Consultant is of a specialist nature, or for some other reason, the Consultant may wish to have part of the service provided by a specialist. The criteria for rejecting a proposed specialist are stated in the contract in general terms, but the contract also makes clear that the Consultant remains responsible for providing the services. Acceptance is in two stages – first the subconsultant is accepted and second the proposed conditions of the subcontract are subject to the Employer's acceptance, unless an NEC contract is to be used or the Employer decides he does not wish to see them. Where the cost reimbursable options C or E is used, the contract data are also subject to the Employer's acceptance.

The adjudicator's role is to decide disputes

The procedure specified in the contract for resolving disputes is that of adjudication. This has been briefly referred to earlier in the discussion on options W1 and W2. Generally, adjudication has made available a quick and relatively cheap means of resolving the majority of disputes that occur in consultancy contracts.

The adjudicator may be appointed by a number of different methods, but the objective should always be to appoint a person who has the confidence of, and is acceptable to, both parties. He should be appointed under a tripartite contract, using the NEC Adjudicator's Contract (AC), and preferably soon after the PSC has come into existence. His services are called upon only when the parties have failed to resolve a dispute. He is required to decide the dispute within 28 days (or longer agreed period) and is paid on a time basis. His costs, under the AC, are shared equally between the parties unless otherwise agreed.

The adjudicator's decision is legally enforceable but a dissatisfied party may take the dispute further to the 'tribunal', which is specified in the contract as either (normally) arbitration or litigation.

The terminology used in the PSC is generally different from that in many traditional consultancy contracts

Since the PSC is very different from other published consultancy contracts, both in principle and detail, there would be a serious risk in using traditional terms of assuming similar meanings. This probably explains why different terms have been used – not merely to introduce change for its own sake. Some of the more important terms are described in the following.

Completion
This is a defined term that seeks to remove, as far as possible, subjective judgement in deciding when the Consultant's work has been completed. Rather than rely on terms such as substantial or practical completion, the definition anticipates that the contract itself should state what completion is by an appropriate statement in the scope (the scope is a detailed description of the services that the Consultant is required to provide). The definition also allows for defects in the completed services, on condition that they do not prevent the Employer from using the services or others from doing their work. If the scope is silent on what work is to be done by the completion date, the test to be applied is whether or not the Employer can use the services and others can do their work. It is in the interests of both parties to establish when the obligation to provide the services has been completed.

Completion date
This is the date by which the Consultant is required to complete his provision of the services. The date may be changed during the contract, under provisions for such matters as compensation events and acceleration. The obligation is a very specific one and contrasts with some of the more traditional consultants' contracts. The latter sometimes limit the obligation to 'reasonable endeavours' to complete, in accordance with a programme to be agreed from time to time, or to work to a timetable but without any warranty that the consultant will complete in accordance with the timetable.

Defect
This definition is expressed in absolute terms in that it is a part of the services that does not comply with the scope or the law. It includes no element of liability, as in the definition in the ECC. Thus it is possible that the Consultant, having complied with his obligations to provide the services using reasonable skill and care, still produces services which are defective. Irrespective of liability, the Consultant is required to correct defects in the services. This

means that the Consultant may be correcting work after having carried out his duties under the contract. This is why such correcting work is a compensation event. It is likely, however, that most defects in the services will be the responsibility of the Consultant.

Key date This term has been introduced into a number of the contracts in the NEC3 family. It is to be distinguished from sectional completion using option X5. A key date is the date by which the services must reach a stated condition. One example would be a contract for design services in which the Consultant must do sufficient design to determine what diversions of public services will be required. This information will then enable the Employer to start negotiations with the authorities owning the services. Another example would be, say, in a contract for design of a building, where the Employer requires the design to be prepared in sufficient detail to enable him to commission a services consultant to design the services for the building.

Key dates may be changed on the instruction of the Employer or under other provisions of the contract.

Risk register This is a further term that has been introduced into a number of contracts in the NEC3 family. The importance of risk in construction has been increasingly recognised in recent years to the extent that, on major contracts, it is common practice to hold risk workshops at an early stage in the development of a project. The main purpose of risk workshops is to identify risks, discuss how to eliminate or reduce them, and then how to manage them by allocating them to one or other of the parties. Risks carried by the Consultant are priced and due allowance is made for them in pricing the contract. In the PSC the contents of the risk register are decided before the contract date in both parts of the contract data, and augmented as further risks are identified as a consequence of the early warning procedure.

Scope This term defines what the Consultant is required to do in providing the services and how he is to do it, as may be appropriate. It is analogous to the works information in the ECC. The scope is contained in a document, which must be very carefully drafted; a less experienced Employer may need to seek assistance in its preparation. In the PSC conditions of contract, there are several references to information contained in the scope. The Employer has authority to change the scope during the course of the contract.

Time charge This term is similar to the Defined Cost of the ECC but with important differences. Basically, it is the product of the number of hours and the staff rates listed in the contract data – the latter including all the Consultant's overheads and profit. It is used for calculating the amount due to the Consultant in the cost reimbursement and target options, C and E, and also in option G. It is also used to assess compensation events for all options.

Activity schedule As in the ECC, this is a pricing document used in the lump sum option A and the target option C. In option A it determines the lump sum price for providing the services. It consists of a list of the Consultant's activities prepared and priced by the Consultant.

The Prices This is a term that has different definitions according to the main option used. It forms the basis on which payments to the Consultant are calculated.

The Price for Service Provided to Date This term has different definitions in the main options. It is used to calculate the periodic payments made to the Consultant throughout the contract period.

The PSC contains some provisions that are different from those in traditional consultancy contracts

Some of the more important and distinguishing features are described in the following.

Communications Effective communication between Employer and Consultant is most important for the smooth running of any consultancy contract. Purely verbal communication is not valid under the PSC. As in other NEC forms of contract, communications between the parties are required to be in a form that can be read, copied or recorded. This does not, of course, prevent the parties from communicating verbally but, to have contractual effect, an instruction, notification or other communication must be in written form. For the purpose of managing communications efficiently, time limits for reply are specified in several clauses of the contract. Where no period is specified in a clause, the standard period for reply stated in the contract data applies. Failure by the Employer to reply within the specified time constitutes a compensation event with possible additional payment to the Consultant where the Consultant's costs and completion times are adversely affected.

Acceptance criteria and procedures are important elements of the PSC. Whenever the Consultant is required to make a submission to the Employer for his acceptance, the criteria for rejection of the submission are also stated. These criteria are stated in objective terms as far as possible, doubtless with the object of avoiding the Consultant being at the mercy of the whim of an unreasonable Employer for any particular submission. This does not, however, prevent the Employer from rejecting a submission for other reasons should he so wish, but in that case rejection is a compensation event. This procedure is clearly intended to minimise misunderstandings, and limit the risk of disputes occurring.

The contract makes clear that the Employer's acceptance does not result in a transfer of risk. Liability of the Consultant remains as stated in the contract.

Early warning The procedure described under this heading is common to other NEC3 forms of contract. It emphasises the benefits of early recognition of potential problems and their early solution – 'attack the problem, not the party'. Most traditional contracts suffer from the absence of an effective way of managing and dealing with problems as they arise. The consequence is the lack of motivation on the part of the parties to co-operate in seeking practical solutions. Rather, each party is motivated to blame the other as far as possible and seek contractual recompense. The early warning procedure in the PSC gives priority to co-operating in solving the problem as soon as possible after it has arisen. The clause requires the Consultant to formally notify the Employer (and vice versa) of any matter that could adversely affect the outcome of the contract. The parties are then encouraged to co-operate in seeking solutions to the problem that has been identified, normally by means of a risk reduction meeting. The four items of the standard agenda for the meeting are set out in the clause, the object of which is to find and agree solutions and decide on the actions to be taken. Liabilities and payment in accordance with the contract will follow from the decisions of the meeting. It will be evident from this that the Employer is much more involved in the Consultant's activities and progress than is the case with most traditional contracts. Thus, if the proper procedures are followed, there should be no late surprises for either party.

Such is the importance of the early warning procedure, that a sanction is included in the event that the Consultant fails to notify an early warning when he could have done so. The sanction comprises a modified assessment of the compensation event, which follows from the matter that is the subject of the early warning.

The programme The contents of the programme which is to be submitted by the Consultant to the Employer for acceptance are set out in detail. They are very similar to the similar requirements in the ECC and include much more than dates and timing. They include such things as method statements on how the Consultant proposes to provide the services, resources, float, time risk allowances and timing of the work of others who may be involved. Thus, in the PSC, in contrast with many traditional consultants' contracts, the Employer is much more involved in how and when the Consultant does his work. This recording of the detail of the Consultant's intentions is likely to facilitate the management of change of circumstances or of the Employers' requirements and, hopefully, minimise the occurrence of misunderstandings and disputes.

The programme is not a contract document. As in most contracts, it is merely a statement of the Consultant's intentions. However, certain parts of the programme acquire contractual status in that departure from some of the programmed dates and times results in compensation events. The Consultant is free to revise the programme as he chooses, but this is always subject to acceptance by the Employer.

Payment The Consultant is required to submit to the Employer an invoice at intervals throughout the duration of the contract. The intervals are stated in the contract data and are usually monthly. The 'amount due' is the total amount due up to the assessment date and the invoice is for the change in the amount due at the assessment date. Calculation of the amount due depends on the main option selected for the contract. It consists of three parts, namely, the Price for Services Provided to Date, expenses and other amounts due in accordance with the contract. The Consultant's expenses are as defined in the contract data. This may vary from 'at cost' to standard allowances specified by the Employer. Items of expenses not so defined are not payable as a separate item and thus must be allowed for in pricing the contract – such as the prices in the activity schedule or staff rates. Where the contract is subject to the Housing Grants, Construction and Regeneration Act 1996, secondary option Y(UK)2 should be included to ensure that the payment provisions comply with the Act. The effect on payment is discussed under the heading of the secondary option Y(UK)2.

Compensation events These are events that may entitle the Consultant to extra payment and time (or in some cases reduced payment). They are thus matters that are at the financial risk of the Employer. There are 12 compensation events listed in the core clauses. A further three are included in main option G and others occur in the secondary options. Several compensation events describe a failure by the Employer to comply with specific provisions of the contract. Also included is a breach of contract by the Employer, not otherwise listed. Probably the most frequently occurring compensation event is the first one listed in the core clauses, namely, a change to the scope instructed by the Employer.

In order to ensure that the contract is managed effectively and efficiently, it is advisable to develop systematic procedures at the start of the contract. The first step in dealing with a compensation event is the formal notifying the other party of the event. Where the event consists of issuing an Employer's instruction, the Employer should, at the same time, notify it as a compensation event and instruct the Consultant to submit quotations. In other cases the Consultant is required to notify the Employer of the event he believes has occurred. The third edition of the PSC has introduced a time bar of eight weeks for notification by the Consultant, dating from when he became aware of the event.

The effect of the quotations to be submitted by the Consultant depends on the main option. For option A, the quotation determines the payment to be made to the Consultant in respect of the compensation event. For option C it affects only the target, and for option E it affects only the forecast of final cost. The effect for option G will vary depending on whether there are relevant items in the schedule and whether the relevant items in the task schedule are lump sums or time based. Qualified quotations are possible but only on the Employer's instructions.

Compensation events are assessed on the basis of the impact the event has on the Consultant's time charge. Once a quotation has been agreed, it cannot be changed except in the case of the qualified quotation where the assumptions made turn out to be incorrect. This early agreement of the effects of change provides both parties with considerable certainty of the final outcome. In assessing compensation events, timing effects as well as the financial effects are always considered. The test to be applied to an event is whether or not the planned completion date is affected. If it is not, no change to the completion date is due. If it is, the completion date is changed accordingly.

Included in the compensation event procedures are sanctions where the Employer fails to adhere to the procedures within the stated times. For instance, failure by the Employer to accept a quotation within the time allowed may result in 'deemed' acceptance after appropriate notice is given by the Consultant.

When to use the PSC

From the above description, it is clear that the PSC can be used for all kinds of consultancy appointments involving the provision of professional services. While it was originally produced to cater for the appointment of consultants in the construction industry, it may also be used for the appointment of consultants in other spheres of professional life. The services of consultants in the construction industry are normally required at various stages through the life of any project, from inception, feasibility and initial appraisal, to outline and detail design, tender and supervision of construction, right through to completion of the project. The consultants may include engineers, architects, designers, quantity surveyors, cost consultants, landscape architects and project managers. The PSC is also suitable for the appointment of the Project Manager and Supervisor under the ECC and for the appointment by a Contractor of a designer under a design and build contract. In the latter case, the PSC may be used as a subcontract; guidance on how the PSC may be adapted for use as a subcontract is given in the published guidance notes on the PSC.

How to use the PSC

Having decided to use the PSC, the Employer's next task is to decide the contract strategy and in particular which main option is most suited to the circumstances of the particular appointment. He will then need to decide which secondary options to incorporate in the contract and draft any z clauses he wishes to add. One of his most important tasks is to write the scope, since this is the document that defines in terms which are as precise as possible, what the task of the Consultant is. In certain situations it may be appropriate to seek the help of the intended Consultant to draft the scope as a joint exercise. The Employer should then complete part 1 of the contract data and prepare any further documents referred to in the contract data.

The next stage will depend on whether the Employer intends to invite tenders from more than one consultant or to invite a consultant whom he has already selected by a separate procurement process. In both cases, tenderers or the selected consultant will be invited to complete part 2 of the contract data, which the Employer will issue as a pro forma. If the Employer already has some form of framework agreement with a number of consultants, he is likely to invite one of them to submit a proposal and negotiate terms in accordance with that agreement. Once a Consultant has been appointed, one of the first tasks of the Consultant is to submit a programme to the Employer, unless one has already been submitted as part of his offer. It is also usual practice to meet the Consultant to discuss the practical details of the contract procedures and, in the case of major appointments, to undertake joint training of the people to be involved in running the contract. An important aspect at this early stage is to develop a team-building strategy to motivate all parties to co-operate in achieving the project objectives.

Exercises

(1) In the following circumstances, decide whether you would use the PSC to appoint a consultant and, if so, what main and secondary options you would recommend, making any necessary assumptions:

(a) A contractor is tendering for a design and build contract to be carried out under the ECC, using main option C, for a major road bridge crossing of a tidal estuary. He wishes to appoint a consultant to investigate three possible designs with estimated costs.

(b) An employer wishes to appoint a consultant as Project Manager and Supervisor together with assistant staff on a £15 m contract under the ECC for the construction of six new sewage treatment works in an African country. The country has a record of high inflation and the area is subject to terrorist attacks from local tribes.

(c) A local authority proposes to provide a business park to encourage businesses to locate in their area to alleviate unemployment. There are three possible sites and the authority wishes to appoint a consultant to advise and report on their relative merits and budget costs of the development.

(d) An employer wishes to appoint a consultant to design an immersed tube tunnel for a new road across a wide river. The consultant will also be required to prepare full detailed drawings and other documents to enable tenders to be invited. The employer also wants the right to use the same design to build a parallel tunnel ten years later, when it is anticipated that the increased traffic volumes will make a further tunnel necessary.

(e) The owner of a large hospital wishes to appoint a consulting structural engineer so that he can call upon his services, as necessary, to advise on structural alterations from time to time.

(f) A pharmaceutical company uses a six-storey building for research purposes. The building is 20 years old but does not meet the requirements for modern research facilities. The options available are to refurbish the building involving major structural alterations, or to demolish and rebuild. The company wishes to appoint a consultant to advise.

7 The Term Service Contract (TSC)

Origins

A consultative edition of the TSC was published in 2002 and the first edition was published in June 2005 together with revised editions of other contracts in the NEC family. A number of organisations had decided to use the consultative version and, therefore, the first edition not only incorporated amendments, which had been made in updating all the other contracts comprising NEC3, but also incorporated results of feed back from its use in practical situations. This contract received the official endorsement of the UK's Office of Government Commerce (OGC) in that it complies fully with the *Achieving Excellence in Construction* (AEC) principles.

Most large organisations, including local authorities, have some form of standard maintenance-type contract. Many of these are based on standard construction contracts, which have been adapted for use, not always success-fully, for maintenance-type work. The publication of the TSC was clearly intended to meet a need, where an employer wished to incorporate the principles and benefits of the NEC system in a contract for the provision of services.

Essential and distinguishing features of the TSC

The TSC is a contract under which a Contractor provides a service of some kind to an Employer over a period of time, called the service period. The service to be provided is defined in the Service Information. It is important to distinguish the TSC from enabling or 'call-off' types of contract, such as the Term version of the ICE Conditions of Contract or the JCT Measured Term Contract. In these forms of contract, there is no precise amount of work at the outset; the Contractor carries out work only when specifically instructed by the Employer or person authorised by the Employer to do so. Effectively it is a schedule of rates contract where the work to be done is priced at rates in the contract. By way of contrast, the TSC defines the work to be done throughout the service period and the Contractor is paid for this work. However, there is provision in one of the secondary options of the TSC for the Service Manager to instruct discrete items of work to be carried out as need arises. This is a practical provision to avoid the need to enter into separate contracts for each small job.

The possible use of the TSC is very wide-ranging, extending from, say, sweeping the pavements in the high street, maintaining public parks or buildings, or refuse disposal, to such major services as maintaining a nuclear power station for a period of, say, ten years. Nor is its use confined to work in the construction sector. Its drafting enables it to be used for providing any kind of service, such as ambulance services for a group of hospitals or providing television programmes in a particular area.

The TSC is different from the PSC in a number of important aspects Whereas the PSC is designed for use where the Employer requires a profes-sional service with its unique requirements, the TSC is used for a service mainly comprising physical work. The end product of a PSC contract may vary from a report of a feasibility study to a set of tender drawings. Thus the majority of work carried out under the PSC is project work, whereas the TSC is for mainte-nance-type work. Normally, work under the TSC is carried out on the Employer's premises, whereas a Consultant appointed under the PSC does his work in his own office or, in the case of site supervision, on a construction site.

Three of the PSC main options (A, C and E) are similar to those of the TSC but the PSC has an additional main option, namely option G, which is described as a Term Option and is a call-off option. The TSC includes this provision but only as a secondary option X19.

The TSC is different in nature from the ECC in many respects The ECC is a contract to 'provide the works', which is part or the whole of a particular project, whereas the TSC is a contract to 'provide the services'. The latter consists essentially of maintaining an existing condition of an Employer's asset over a period of time (the service period) to permit the Employer, or in some cases the public, to continue to use a facility. It would not, in general, include changing or improving the Employer's asset – that would be a project. However, a certain amount of improving the condition of an asset (sometimes called betterment) may be inevitable or otherwise reasonably included in a TSC. For the purposes of the TSC, maintenance includes renewal or replacement of items which have worn out or reached the end of their useful lives.

The equivalent of the ECC's Project Manager in the TSC is the Service Manager – he manages the contract on behalf of the Employer. There is, however, no equivalent of the ECC's Supervisor in the TSC.

The published TSC document

As in the case of other NEC forms of contract, the published TSC document is not a contract but rather a number of statements and provisions from which a contract to suit particular circumstances may be prepared. An informed Employer should have little difficulty in preparing the tender or contract documents for work to be carried out under the TSC. However, a less experienced Employer may require professional assistance and achieve this by appointing a consultant to advise.

A contract under the TSC will comprise:

(a) Core clauses. These must be included in every contract.
(b) One main option selected from the three main options, A, C and E. Selection of the main option determines the allocation of risk between the Employer and Contractor and how the Employer is to pay the Contractor. The numbering of the clauses follows on from the core clauses and under the same side headings as in the core clauses.
(c) A dispute resolution option W1 or W2 according to whether the UK's Housing Grants, Construction and Regeneration Act 1996 applies to the contract.
(d) Secondary options selected from the numbered options X1 to X4, X12, X13, X17 to X20, and the two options numbered Y(UK)2 and Y(UK)3. This selection determines the allocation of further risks.
(e) Additional bespoke conditions of contract indicated by the letter z. These are normally decided by the Employer or by agreement between the parties. Because of the options available, the need for additional conditions should be minimal.
(f) Contract data. This is in two parts: part 1 being prepared by the Employer and part 2 prepared in part by the Employer for completion by tendering contractors or by a contractor selected under a framework or similar agreement. The contract data contains information specific to the contract. Some information in the contract data takes the form of references to other documents, which are thereby incorporated into the contract.

One of the most important decisions the Employer has to make is to decide which main option to use

There are three main options: A, C and E. The Contractor carries the greatest financial risk under option A and the least under option E. Selection is determined by the nature of the service to be provided and how the risks are to be dealt with.

The lump sum option. Option A: Priced contract with price list

This is very similar to option A of the ECC and PSC except that the activity schedule is replaced by the price list. It is a lump sum contract in which the Contractor undertakes to provide the services described in the contract for a sum of money. He carries the financial risk of providing the services for the lump sum. The price list is similar in form and function to the price list in the ECSC, except that it is not included as part of the published standard contract documents. There is, however, a pro forma price list in an appendix to the guidance notes. The columns are similar to those of a traditional bill of quantities but there is no method of measurement to determine what the items should be and how they are to be measured. But it is possible that the Employer may wish to specify rules under which a tenderer is required to draft and price the price list. This means that, although the quantity of an item may have to be corrected according to the work actually done by the Contractor, the number of items will not change other than adding items as a result of a compensation event or the issue of a task order under option X19 (see later).

The items in the price list are of two kinds – lump sum items and remeasurable items. The lump sum amounts are paid when the Contractor has completed the work described in the item; there is no provision for payment of a proportion of an item. Payment for remeasurable items is made according to the actual quantity of work that the Contractor has carried out.

Where a particular service is to be provided on a regular basis throughout the service period, say daily or weekly, the Contractor may be paid a regular (say monthly) amount. Under this payment arrangement, the unit entered in the price list would be, say 'month' and the expected quantity would be the total number of months for which the service is to be provided.

The price list may be compiled by either the Employer or the Contractor but the pricing is always done by the Contractor. Careful thought needs to be given by the Contractor to compiling and pricing of the price list; he needs to ensure that the number and descriptions of the items in the price list are such as will maintain an acceptable cash flow. In pricing each item, the Contractor takes responsibility for estimating and pricing the resources he is likely to need, making due allowance for risk, overheads and profit. One of the greatest advantages of an option A contract is its straightforward administration.

Where the service can be precisely defined, option A is the option most likely to be selected by the Employer.

The target option with the target established as a combination of lump sums and remeasurable items. Option C: Target contract with price list

Under this option, the Contractor tenders or negotiates a target price (defined as the Prices) using a price list. During the course of the contract, the Contractor is paid his actual costs as defined in the contract under the term 'Defined Cost' plus the Fee. The term 'Fee' broadly covers the Contractor's overheads and profit and any other cost item not included under Defined Cost. The Fee is calculated by applying the tendered fee percentages in contract data part 2 to Defined Cost. At specific times during the contract (unlikely to be at less than yearly intervals) the Defined Cost plus Fee is compared with the target at that time. If this shows that a saving has been made, in that Defined Cost plus Fee is less than the target, the difference is shared between the Employer and the Contractor in proportions that are defined in the contract. If it shows that the target has been exceeded, the Contractor must pay back a proportion of the excess. During the course of the contract, the target price is adjusted to cater for compensation events, remeasurement and task orders (where option X19 is included).

Thus, option C is basically a cost reimbursement contract, which incorporates an incentive for the Contractor to minimise costs. Savings and over-runs are shared between the parties. It is most likely to be used in providing services which involve high risk for the Contractor such that a priced contract under option A could create serious problems in knowing how to price the risks. Pricing risks at too high a level could jeopardise the Contractor's chances of being appointed. On the other hand, under-pricing risk could create serious difficulties for the Contractor if and when the risks eventuate. It is for this reason that the sharing of risks in such circumstances is likely to reduce the occurrence of disputes and produce a more satisfactory result for both parties. Use of option C is therefore appropriate where a lump sum contract, as in option A, would place too much risk on the Contractor but where a purely cost reimbursable contract would place too much risk on the Employer and thus provide little incentive for the Contractor to work efficiently and economically.

The option where the Contractor is paid his costs. Option E: Cost reimbursable contract

Under this option the Contractor takes a very small risk since he is paid his actual cost (Defined Cost, which includes a deduction for certain Disallowed Cost as defined) plus the Fee with only a small number of constraints to protect the Employer from the effects of incompetence and inefficient working on the part of the Contractor. It is most likely to be used where the work to be carried out in providing the services cannot be easily defined at the outset of the contract and when the risks in doing the work are great. The option may also be suitable for providing services, which are of an experimental nature or for research work where decisions on what is required may have to be made on a day-to-day basis. But such uses are not likely to be common. The price list in an option E contract is used only for forecasting purposes.

The dispute resolution options determine the procedures for dealing with disputes

The dispute resolution options W1 and W2

As in other NEC contracts, the parties are required to act in a spirit of mutual trust and co-operation as a contractual obligation. But, in spite of this, disputes do arise for a variety of reasons. It has been the practice for many years to include in conditions of contract, procedures for resolving disputes without recourse to the courts. The method of dispute resolution specified in the past, has usually been arbitration. However the NEC introduced adjudication as the initial method of dealing with disputes. In the UK this has been made a statutory right of the parties to a 'construction contract' by virtue of the Housing Grants, Construction and Regeneration Act 1996 (the Act).

The TSC has introduced two separate dispute resolution options:

- Option W1 for contracts not subject to the Act.
- Option W2 for contracts which are subject to the Act.

The main provisions of these options have been described in the chapter on the ECC.

As defined in the Act, a construction contract means an agreement for the carrying out of construction operations. The Act includes a definition of construction operations as well as a definition of certain activities that are not construction operations. Construction operations are defined very widely. They include, for instance, repair and maintenance of buildings or structures forming part of the land, whether permanent or not. They also include repair and maintenance of any works forming part of the land, including walls, roadworks, power-lines, telecommunication apparatus, aircraft runways, docks and harbours, railways, inland waterways, pipelines, reservoirs, water-mains, sewers, industrial plant and installations for the purposes of land drainage, coast protection or defence, and painting or decorating the internal or external surfaces of any building or structure. It is clear from this that many TSC contracts in the UK will be subject to the Act but, where there is doubt, the Act should be consulted.

The adjudicator is named in the contract data and should be a person with some knowledge and experience of the kind of work that is the subject of the contract in question.

Various further risks need to be considered to decide how they can be reduced and which party should carry them. This is the function of secondary options

Which party should carry the risk of price increases during the period of the contract? Option X1: Price adjustment for inflation

At most times, for a two-year contract in the UK, the inflation risk is not great. Hence, it would not be unreasonable to require a Contractor to take that risk and price the contract accordingly. In doing so he would make due allowance in his pricing for price and wage increases throughout the period of the contract.

But for, say, a ten-year maintenance contract, the risk is considerable and, therefore, it is in the interest of both parties that the Employer should take the risk. Whether or not to include this option is a matter of judgement in all the circumstances.

The method of calculating adjustments for inflation is by means of a formula, which makes use of price indices for labour and materials. These indices are published nationally in many countries. The particular indices selected should be those that reflect the nature and content of the services. In countries where no such indices are published, it would not be possible to use this option; an alternative provision would be required to ensure that the Contractor is reimbursed appropriately for changes in prices. At each date when the payment due to the Contractor is assessed, a price adjustment factor is calculated from the base index and the latest index. This is then multiplied by the change in the amount due since the previous assessment date.

In the case of option A, the adjustment is made to the amount otherwise due to the Contractor as calculated from the price list. For option C, adjustment is made only to the target, as the Contractor is paid actual cost (Defined Cost plus Fee) periodically throughout the contract and therefore recovers any price increases that may have occurred. In option E, the Employer automatically carries the risk of inflation.

Which party should pay for the consequences of any change in the law? Option X2: Changes in the law

During the service period, a change in the law may affect the Contractor's costs. The longer the service period the greater the risk. In some countries, such a change, for example a change in import duties or customs payments, can have a major effect on the Contractor's costs. Inclusion of this option transfers the risk of the effects of changes in the law from the Contractor to the Employer. It applies whether the changes in the law have the effect of increasing or reducing the Contractor's costs. The compensation event procedure is used to implement this option.

If the Contractor is to be paid in more than one currency, how is this done and which party takes the risk of changes in the exchange rates? Option X3: Multiple currencies

This option is intended for use on international contracts where the Contractor may require to be paid in more than one currency. A 'currency of this contract' is established for each contract and stated in contract data part 1. This option is designed for use only with option A. The Contractor prices his offer in the 'currency of this contract' and the items of work that are to be paid for in a different currency are listed in the contract data. Conversion to other currencies is done by using the 'exchange rates' stated in the contract data. These are fixed at a definite date (usually some weeks before the contract date), which is also stated in the contract data. In option C the Contractor is paid on the basis of his actual cost. Thus, payment is made in whatever currency the Contractor incurs the cost. To calculate the Contractor's share, payments are converted to the 'currency of this contract' using the 'exchange rates'.

What security does the Employer require in the event of the Contractor's failure to carry out his obligations? Option X4: Parent company guarantee

Where the Contractor is owned by a parent group of companies, the inclusion of this option gives the Employer greater security of the Contractor's performance in the form of a guarantee by the parent company. The form of guarantee is stated in the contract. Failure by the Contractor to provide the guarantee within stated time periods entitles the Employer to terminate the contract.

How can the Contractor be encouraged to work with other parties as a team, to produce a successful outcome? Option X12: Partnering

One of the main intentions of the NEC provisions is to encourage collaborative relationships between the contracting parties rather than to create an environment of confrontation as is the case with many traditional forms of contract. Partnering has been introduced in recent years to encourage collaboration in a similar way but between all parties who may be involved in a particular project or series of projects. When partnering was first introduced, the arrangement between the various parties was non-contractual. However, there arose a demand for such arrangements to be enshrined in a legally binding contract or contracts. To many, this seemed to be an attempt to combine obligations that were contradictory, but this view may have resulted from the confrontational nature of traditional contracts. However, in response to the demand, the Construction Industry Council (CIC) published a 'Guide to Project Team

Partnering'. This was not a partnering contract but merely provided guidance on matters that were considered to be essential to a partnering contract.

Since the NEC family of contracts is based on a policy of co-operation rather than confrontation, all that was necessary to create an effective partnership was to devise a mechanism for creating the necessary relationships between the various parties already operating under one of the NEC contracts. This is the function and effect of option X12, which is described in the chapter on the ECC. This option could be used where the Employer has a contract or contracts under the ECC or ECSC or PSC at the same time and in the same area as a contract under the TSC.

How can the Employer obtain some financial security when the Contractor fails to perform? Option X13: Performance bond

In most construction contracts, the Employer requires some form of security as a protection against a Contractor's failure to perform his obligations for whatever reason. Under the TSC this takes the form of a performance bond provided by a bank or insurance company. In effect this is a method of transferring the risk of a Contractor's non-performance to a third party. Further details of this option are given in the chapter on the ECC.

What price should the Contractor pay if he fails to carry out his obligations? Option X17: Low performance damages

This option provides for payment by the Contractor of damages at levels agreed in the contract, to reflect a failure by the Contractor to reach specified levels of performance. Under English law, penalties are not enforceable. Hence, the amount of damages stated in the contract should not be greater than a genuine estimate of the financial damage suffered by the Employer as a result of the reduced performance. In effect, low performance damages are the opposite of key performance indicators. Under the latter, the Contractor is rewarded for achieving specified targets. Examples of low performance damages in a TSC are failure to respond within a stated time to an instruction to carry out responsive repairs to a property; failure to complete discrete items of work in the times specified in the contract and failure as a result of a post inspection.

Should there be some financial limit to the Contractor's legal liability? Option X18: Limitation of liability

This option permits the fixing of limits to various liabilities which the Contractor may have under the contract, or even outside the contract in the case of his total liability. The Contractor's liability may be considerable, particularly in the TSC where the Contractor may be working on or near to an asset belonging to the Employer. The potential liability in such cases may be out of all proportion to the value of the contract – hence the need, for practical and commercial reasons, to limit the Contractor's liability. The separate liabilities provided for are indirect or consequential loss, damage to the Employer's property and the consequences of defects due to the Contractor's design of an item of equipment. In addition, there is provision for limiting the Contractor's total liability for all matters. In many jurisdictions there is a statutory limit to the time after which the parties' obligations under a contract cease (the limitation period). Where there is no such statutory limit or where the parties wish to change the limitation period, the final clause in this option can be used to introduce an 'end of liability date'.

An Employer may wish to call upon the Contractor to carry out discrete items of work not included in the contract, the need for which becomes evident only during the service period. Option X19: Task Order

The inclusion of this option fulfils a practical purpose, namely to avoid the need to execute separate contracts for relatively small items of additional work which the Employer may wish to have carried out. It is an enabling or 'call-off' option, which can be used with all main options. The task instructed to be carried out is described in a 'task order', which also includes the start and completion dates, delay damages to be applied in the event of delay to completion of the task, and an assessment of the price for the work compiled from the price list. A task of this nature can be regarded as a 'mini project'. The Contractor is then required to submit to the Service Manager a detailed programme showing how and when the task is to be carried out, and what resources are proposed.

Should the Contractor be rewarded for meeting specified targets? Option X 20: Key Performance Indicators

An important aspect of recent commercial and industrial life has been the introduction of targets and incentives with the object of achieving continuous improvement. The aim of Key Performance Indicators (KPIs) is to provide the Contractor with an incentive to achieve the Employer's objectives. Further details of this option are given in the chapter on the ECC. Examples of KPIs, which may be used in a TSC, are number of notified defects, number of days to complete specific items of work, number of complaints by the public made against the Contractor.

What is necessary to make the contract comply with UK legal requirements? Option Y(UK)2: The Housing Grants, Construction and Regeneration Act 1996.

The construction part of this Act was introduced to deal with two matters, namely adjudication of disputes and payment. The first of these is dealt with by selecting the appropriate dispute resolution option W2 and payment is dealt with by incorporating this option Y(UK)2. It should be used only where the Act applies to the work in question. Further details of this option are given in the chapter on the ECC

Option Y(UK)3: The Contracts (Rights of Third Parties) Act 1999

This Act permits third parties to exercise rights under a contract even though they are not parties to that contract. In the TSC, provision has been made for specific rights to be enforced by people or organisations stated in the contract. This may be particularly useful where the parties wish to give certain rights to others such as partners under option X12.

Is any fine-tuning required to meet the requirements of the Employer and the Contractor and to suit the circumstances of the project? Option Z: Additional conditions of contract

In spite of the flexibility available in the structure of the TSC, it will often be necessary to make additional provisions to suit the circumstances of a particular project – the fine-tuning. Such additional z clauses comprise the bespoke element of the contract and are usually decided by the Employer but sometimes negotiated between Employer and Contractor. Additional clauses should not be used to substantially change the overall balance of risk between the parties as this may cause conflict within the contract. Additional clauses should be carefully drafted to harmonise with the standard clauses avoiding overlap wherever possible.

The roles, duties and powers of the Employer, the Contractor and the Service Manager are described in detail throughout the conditions of contract

The contracting parties are the Employer and the Contractor, each of which has specific powers and duties described in the contract. In addition, other parties are referred to. The Service Manager, for instance, is appointed by, and operates on behalf of, the Employer. Subcontractors are defined in the contract to distinguish them from suppliers. The Adjudicator is named in part 1 of the contract data. His function is to decide disputes referred to him by one of the contracting parties, under option W1 or W2. 'Others' is also a defined term. Designers, as such, are not expressly mentioned in the contract; in a contract under the TSC their main function would be to design temporary works.

The roles, duties and powers of these various parties are described throughout the conditions. However, section 2 of the conditions covers the main responsibilities of the Contractor. The kind of people appointed to these positions will depend on a number of factors, the most important of which is the nature and extent of the work to be carried out. Thus, the person appointed as Service Manager for maintenance of a large building complex may well be an architect, or for a large engineering installation, a civil, mechanical or chemical engineer.

The basic duty of most of the above parties is stated in the first clause of the contract. Presumably it is placed at the beginning because it underpins the whole contract. It describes not only what the parties must do but also how they are to act. The first of these is inherent in the nature of contracts. The various actions are generally described throughout the conditions in the present tense and the use of the word 'shall' makes the actions mandatory. The second duty expressly states that the parties are to act in a spirit of mutual trust and co-operation. It is not usual for standard contracts to state how parties are to carry out their duties, even though the duty of co-operation has sometimes been implied. This requirement follows the recommendation of the Latham Report and has been the subject of much legal comment, particularly in relation to how it can be enforced. But its inclusion is largely pragmatic, for the benefit of the parties and in the interests of the service to be provided, with the aim of avoiding disputes and litigation.

There is very little mention of the Employer in the contract, but he has some important basic rights and responsibilities

The Service Manager exercises most of the day-to-day management and administrative duties and powers on behalf of the Employer. However, there are a number of basic matters, which are reserved exclusively for the Employer in his role as one of the contracting parties. These include the duty to allow the Contractor access and to provide stated facilities and materials, insurance, termination and the right to refer disputes to adjudication.

The Contractor has many fundamental obligations

While the Contractor's main responsibilities are described in section 2 of the conditions, many other responsibilities are contained in the rest of the conditions. His main obligation is to provide the service and what and how he is to provide it is covered in detail in the Service Information. The Contractor is normally responsible for designing any necessary temporary works or equipment but the Service Manager is entitled to have details of these should he so wish.

Before the Contractor subcontracts any of his work, he is required to seek the acceptance of the Service Manager for the subcontractor and the conditions of the subcontract. This goes further than many standard conditions and illustrates the greater involvement of the Employer's Service Manager in the contract. The Service Manager's involvement is increased still further in the cost reimbursement options C and E, under which the Contractor is required to submit to him the contract data of the subcontract.

The Service Manager occupies a crucial role in the successful managing of the contract

The Service Manager is appointed by the Employer and is named in the contract. His role is crucial in the management of the contract. He has extensive powers and duties in the contract and it is important that he has the requisite qualifications to carry these out. The extent to which the Employer may restrict or qualify these powers and duties is a matter between the Employer and the Service Manager. But the Contractor may assume that the Service Manager has the necessary powers to fulfil his role in administering the contract. The contract contains remedies for the Contractor in the event that the Service Manager fails to act within the times stated in the contract. Thus, it is incumbent on both Employer and Service Manager to ensure that any such restrictions do not prevent the Service Manager from functioning effectively.

The Service Manager should be a person with business experience capable of exercising commercial judgement, particularly in decisions concerning issuing task orders, payments to the Contractor and in assessing the effects of change. On major contracts it would be normal for the Service Manager to delegate some of his powers and duties to others who may be better qualified to fulfil parts of this role. The authority to delegate is unrestricted in the contract.

The Service Manager also has important powers of inspecting and checking the Contractor's work and materials, as described in section 4. This section also includes a mechanism for accepting defects by agreement. This is an extremely useful procedure in cases where, although technically work is non-compliant, its correction is not strictly necessary or even possible. Another important role of the Service Manager is that of certifier. He is required to issue certificates in relation to both payment and termination. In both of these, skilled judgement is required.

The appointment of subcontractors and certain terms of subcontracts are subject to acceptance by the Service Manager

The Contractor may subcontract parts of the services, but that does not affect in any way his basic responsibility to provide the service. Because of the need to distinguish between suppliers (not defined as such in the conditions) and subcontractors, the latter are defined in a specific way. There are three parts to the definition. The third part requires that, to qualify as a subcontractor, a supplier of plant and materials must have carried out some design that is specific in relation to the service. In other words, the definition excludes standard 'off-the-shelf' items.

The clauses dealing with subcontracting, encourage the Contractor to use a NEC contract for the subcontract. Suitable NEC contracts would be the subcontract (ECS) or the short subcontract (ECSS). While use of a NEC contract is not mandatory, the benefits to the Contractor are obvious in that the two contracts are almost back-to-back, leaving little residual risk with the Contractor. The main option in the subcontract may well differ from that of the main contract, and similarly the secondary options, for a number of reasons, may be different. But the requirement of 'mutual trust and co-operation' is effectively mandatory in the sense that its absence provides a valid ground for rejection of a sub-contract proposal. The intention here is clearly to ensure that this fundamental condition extends down the whole of the supply chain.

There is no provision in the contract for nomination of subcontractors.

The Adjudicator's role is to decide disputes

The procedure specified in the contract for resolving disputes, is that of adjudication. Since its inclusion in the first edition of the NEC, the right to have disputes in a construction contract decided by an adjudicator has been enshrined in statute in the UK's Housing Grants, Construction and Regeneration Act 1996. Generally adjudication has made available a quick and relatively cheap means of resolving the majority of disputes that occur in construction contracts.

Further details of adjudication are given in the chapter on the ECC.

The terminology used in the TSC is generally different from that in many traditional contracts

Some of the more important terms are described in the following.

Affected Property   This term replaces the 'site' of the ECC and describes the areas affected by the Contractor's activities. In many TSC contracts this would be property owned or occupied by the Employer. Its main importance is in relation to compensation events and insurance. A change to the Affected Property may adversely affect the Contractor in his providing of the service in terms of cost and also his plan.

Defined Cost   This broadly represents the Contractor's actual cost and in the TSC has two uses. First, for the cost reimbursable contracts using option C or E, it is used to calculate the periodic payments to the Contractor. Second, it is used for assessing compensation events for all main options in cases where there is not an appropriate rate or price in the price list. Assessment in those cases is based on the impact of the compensation event on the Contractor's costs as defined.

Disallowed Cost   This is a long definition and consists of certain Contractor's costs, which are deducted in calculating Defined Cost.

Equipment   This is anything of a temporary nature provided by the Contractor to enable him to do his work. It does not include equipment provided by the Employer for the Contractor's use, or permanent plant and equipment, which is part of the affected property.

Plant and Materials   These are items of a permanent nature supplied by the Contractor as part of his providing the service. In a straightforward maintenance contract it includes replacement items.

Service Information   This information is contained in the documents stated in the contract data. It describes, in detail, what the Contractor is required to do and may include how he is to do it. It normally consists of specifications and may include drawings.

The Prices   This is a term that has different definitions according to the main option used. The Prices all relate to the Price List.

The Price for Services Provided to Date   This term has different definitions according to the main option used. It is used to calculate the periodic payments made to the Contractor progressively throughout the service period.

The TSC contains some provisions that are different from those in more traditional contracts and which characterise its distinguishing features

Some of the more important of these features are described in the following.

Communications Clause 13 recognises the importance of effective communication between all parties in managing the contract efficiently. It does not permit purely verbal communication as a valid means of communication under the contract; it requires that communications are in a form that can be 'read, copied and recorded'. This does not, of course, prevent parties communicating with each other verbally but, to have contractual effect, any instruction or other communication must be in written form. The clause does not allow for confirmation of verbal instructions. The clause requires parties to reply to communications within definite periods of time. Failure by the Service Manager to reply within these times constitutes a compensation event.

Acceptance procedures are also covered in clause 13. Throughout the conditions, whenever the Contractor is to make a submission to the Service Manager for his acceptance, the criteria for rejection of the submission are also stated. These criteria are stated in objective terms as far as possible. This does not prevent the Service Manager from rejecting a submission for other reasons, should he so wish, but in that case rejection is a compensation event. This minimising of purely subjective grounds for rejection of a contractor's submission, is likely to reduce the risk of disputes.

Early warning This has been described by one construction lawyer as the jewel in the crown of the NEC. The procedure described in clause 16 gives priority to co-operating in solving a problem as soon as possible after it arises. Procedures in many traditional contracts have the effect of motivating the parties to adopt a contractual and defensive stance when such problems arise, with both parties reserving their position. This clause requires the Contractor formally to notify the Service Manager (and vice versa) of any matter which could affect the outcome of the contract. The parties are then encouraged to co-operate in seeking solutions to the problem that has been identified, normally by means of a meeting, now called a risk reduction meeting. The agenda for the meeting is set out in the clause, the objective of the meeting being to discuss the problem, find and agree solutions, and decide on the actions to be taken. Liabilities and payment under the contract will follow from the decisions of the meeting.

The Contractor's plan The Contractor's plan is a very important document in the TSC. Because of the range of options available, the Service Manager is much more involved in the Contractor's activities – not only in what he does but also in how he does it. Clause 21 allows for a tender plan, which becomes the first plan on acceptance of the tender, and also a plan submitted by the Contractor after the start of the contract. The importance of the plan is demonstrated by the long list of matters that are required to be included in the plan. Apart from various dates, the list includes statements of methods that the Contractor proposes to use to provide the services together with the resources which he plans to use. It also includes time allowances for matters that are at the Contractor's risk, and timing of the work of the Employer and others. Such detail is of considerable benefit to both parties when it becomes necessary to assess the impact of change. While the plan is no more than a statement of the Contractor's intentions and proposals, certain dates on the plan acquire contractual status. This is clear from the list of compensation events in section 6 of the conditions, three of which refer to dates on the accepted plan. One of these compensation events also makes clear that in the event of one of the 'Others' not performing in accordance with the programme, that risk is taken by the Employer.

Managing defects The procedure for correcting defects is less prescriptive than in the case of the ECC. The Contractor's obligation is to correct defects within a time 'which minimises the adverse effect on the Employer or others'. In recognition of the fact that defects may vary in nature and extent according to circumstances, the TSC does not prescribe a fixed defect correction period. The provision also recognises the fact that the Employer or others may be in continuous occupation of the affected property, and that correction must be done at the Employer's convenience. However, occasionally conditions may be such that it becomes impossible to correct defective work. Circumstances may occur also where the cost of correcting a defect may be great and its correction may be technically, though not contractually, unnecessary. Clause 43 provides a mechanism to deal with such a situation, which is not uncommon in many contracts. Under this clause, a defect can be accepted by agreement of the parties, with some reduction in the prices.

Compensation events

These are events which may entitle the Contractor to extra payment (or in some cases reduced payment). They are thus matters which are at the financial risk of the Employer. There are 14 compensation events listed in the core clauses of section 6 of the conditions. Additional compensation events are included in some of the main and secondary options. The procedures for notifying them are clearly stated in clause 61 and assessment of their financial effects are detailed in clause 63. There is considerable flexibility provided in clause 62.1 whereby the Contractor may submit alternative quotations based on different methods of dealing with a compensation event. Thus the Service Manager in these circumstances may have several priced options and he is free to choose whichever suits him best. The financial aspect of a quotation for a compensation event is calculated by using relevant rates in the Price List. Where there are no such rates, a compensation event is assessed by judging the impact of the event on the Contractor's costs (Defined Cost), Where agreement cannot be reached on the assessment of a compensation event, the fall back position is that the Service Manager decides. If the Contractor does not agree with the Service Manager's assessment, the Contractor's recourse is to refer the dispute to the Adjudicator.

Insurance

There are five categories of insurance which the Contractor is required to provide; this is one more than the ECC insurance requirements. The additional one is insurance of the Employer's property. Since work is to be done on the affected property (which would usually be in the ownership or occupancy of the Employer), the risk of damage in a TSC contract may be much greater than in the case of the ECC. However, if the property concerned is already insured by the Employer, it may not be necessary to require the Contractor to insure it. The contract data allow for changes to the core insurance requirements.

The TSC can be used as a subcontract if some amendments are made

In the absence of a published Term Service Subcontract, it is necessary to adapt the TSC for use as a subcontract for work that the Contractor wishes to subcontract. Guidance on the changes necessary to achieve this is given in an appendix to the published guidance notes. This ensures that the two contracts are back-to-back as far as possible. However, for good reason, the main option in the subcontract may differ from that in the main TSC. The main changes required to the TSC for use as a subcontract are in terminology, the content of the Service Information, requirements for the subcontractor's plan, payment, the timing included in the procedures for compensation events, 'joinder' of disputes and insurance. As mentioned earlier, and as an alternative, the Contractor may elect to use one of the other NEC contracts for subcontract work under the TSC.

When to use the TSC

From the above description, it is clear that the TSC can be used for all kinds of maintenance-type contracts required by a promoter or client (the Employer). Proper management of such contracts inevitably means that the procedures stated in the contract should be meticulously followed.

The TSC should not be used for the provision of professional services (where the NEC Professional Services Contract would be appropriate), or for project-type work (where the ECC or ECSC would be appropriate). The published guidance notes give a number of examples of use of the TSC both within and outwith the construction sector of industry.

How to use the TSC

Having decided to use the TSC, the Employer's next task is to decide his contract strategy and in particular which main option he proposes to use. He will then need to decide which secondary options to incorporate in the contract and draft any z clauses he wishes to include. One of his most time-consuming tasks is to prepare the Service Information. The effort and resources required for this task may largely depend on the design work, if any, for the equipment he intends to use and the amount and complexity of the planning required. He should then complete part 1 of the contract data and any further documents such as the affected property, which are referred to in the contract data.

The next stage will depend on whether the Employer intends to invite competitive tenders, in which case he will prepare a list of tenderers and instructions for tendering. For this purpose, the tender documents will include part 2 of the contract data as a pro-forma, which tenderers will be invited to complete. Assessing competitive tenders will normally include careful scrutiny of the Contractor's plan if submitted as part of the tender. Assessing the financial aspect of tenders will include scrutiny of the contents of the price list and of the fee percentages in part 2 of the contract data. If the Employer already has some form of framework agreement with a number of contractors, he will invite one of them to submit a proposal and negotiate the terms in accordance with that agreement.

Once a Contractor has been appointed, one of the first tasks of the Employer's Service Manager is to decide which powers, if any, he wishes to delegate and to whom. It is also usual practice to meet the Contractor to discuss practical details of the contract procedures. An important aspect at this early stage is to develop a team-building strategy to motivate all parties to co-operate in achieving the contract objectives.

Exercises

(1) Decide whether you would use the TSC for work described in the following, making any necessary assumptions.

   (a) The edges of a narrow rural road are deteriorating because of vehicles using the side verges in attempting to pass. The local authority decides to reconstruct a 1-metre strip on each side and include kerbs and positive drainage. It is to be financed from its maintenance budget.

   (b) A local authority wishes to renew the deteriorating sewerage system in a housing estate, as part of its long-term renewal programme. It is to be financed from its maintenance budget.

   (c) The landscape area of a business park has been neglected due to lack of maintenance and is in urgent need of restoration. The initial work needed is weed removal, tree surgery, pruning of trees and shrubs, replacing dead plants and fertilising. Thereafter, the owner wishes to maintain it properly on a regular basis for the following five years.

   (d) A harbour authority wishes to increase the depth of water alongside its fish quay by 1 metre. It also wants the appointed contractor to do maintenance dredging throughout the harbour for a five-year period to provide a minimum depth of water of 3 metres at Low Water Spring Tides.

   (e) A canal authority wishes to appoint a maintenance contractor to maintain a 60 km length of canal. The work required includes dredging, repair and maintenance of lock gates, sluice valves and pumps, pointing of brickwork in tunnels and retaining walls, and repair and maintenance of tow paths.

   (f) A railway authority intends to let a contract for the maintenance of its boundary fences and 30 railway stations for a period of eight years.

   (g) The owner of a 12-storey office building with a reinforced concrete frame and located in an urban area, wishes to demolish it, level the site and sell it for housing development.

   (h) The walls of a house have suffered extensive structural cracking due to lack of maintenance and settlement caused by adjacent vegetation. The owner has been advised to have the exterior walls underpinned to depths of up to 1 metre.

(2) Decide the main and secondary options you would select for contracts under the TSC in the following situations, making any necessary assumptions.

   (a) A five-year contract for maintaining and operating the public parks for a local authority. The work includes:

      • clearing and disposing of litter
      • pruning and maintaining shrubs trees and flower beds
      • grass cutting and clearing autumn leaves
      • maintaining tennis courts, bowling greens and children's play areas.

   (b) Maintaining 80 km of motorway for five years, including repair of accident damage and a rapid response facility.

   (c) Maintaining and servicing a large hospital for five years. Work includes:

      • maintaining heating, ventilation and lighting
      • painting and decorating on a planned maintenance basis
      • security services
      • providing catering services using the hospital's kitchen and equipment
      • managing the hospital car park.

(d) Maintenance dredging for a harbour authority for ten years.
(e) Providing ambulance services for a group of hospitals for two years.
(f) Winter gritting and snow clearing for all county roads, for a county highway authority. Contract period is five years.
(g) Maintenance and management of a 60 km length of canal for three years. The work required includes:

- dredging and clearing of vegetation
- maintaining and repairing locks, tunnels and towpaths
- managing the leisure facilities
- maintaining water flows
- maintaining and repairing culverts bridges and aqueducts
- dealing with leakages and settlement of embankments.

8 The Adjudicator's Contract (AC)

The method of resolving disputes in all the NEC contracts is adjudication. This is the case whether or not the contract is subject to the Housing Grants, Construction and Regeneration Act 1996. Each contract requires the parties to the contract to appoint and enter into a contract with an Adjudicator under the NEC Adjudicator's Contract. The procedures and timing for appointing the Adjudicator have not been included in the published NEC documentation. However, there is provision for the Adjudicator to be named in the tender documents. This suggests that the Employer decides who the adjudicator should be but, if the Employer decides this unilaterally, there may be difficulty in agreeing the terms of, and executing the contract with, the Adjudicator. The published guidance notes for the NEC contracts suggest various ways in which a suitable Adjudicator agreeable to both parties may be appointed. The parties should then enter into a contract with the Adjudicator incorporating the agreed terms, soon after the contract date of the contract between the main parties. For the purposes of this chapter, the contract between the two parties is called the principal contract. The benefit of executing the Adjudicator's Contract (AC) as soon as possible is that if and when a dispute occurs, unnecessary delay in searching for and appointing a suitable person, is avoided. A disadvantage of delaying the appointment of the Adjudicator is that, in the absence of agreement, it may be necessary to call upon the services of a third party (an adjudicator nominating body) to make the appointment. If this occurs, the third party will usually appoint without reference to the parties.

The AC is, therefore, different from the contracts described in other chapters in that it has no meaningful existence without a principal contract. To that extent, the parties do not have to choose this contract as its use is mandatory. However, this chapter is included for the sake of completeness and to inform the parties using an NEC contract about the nature of the contract they will be required to enter into with the Adjudicator.

Origins

When the first edition of the NEC was published in 1993, the benefits of having standard conditions of contract for the appointment of the Adjudicator soon became apparent. Accordingly, a first edition was published in 1994. This was followed by a second edition in 1998. This sought to make consistent the adjudication provisions of the various NEC contracts and the AC. Detailed adjudication procedures were removed from the AC to avoid conflict with the procedures already contained in the principal contracts. The third edition of the AC was published in 2005 as part of the NEC contracts published under the general title NEC3.

The essential features of the AC

The AC consists of three parts, namely a form of agreement, conditions of contract and contract data.

The form of agreement This simple agreement records the names of the three parties, the fact that the parties to the principal contract have appointed the Adjudicator, and acceptance by the Adjudicator of the appointment under the AC conditions of contract. All three parties sign it.

The conditions of contract The first clause of the AC imports by reference, the conditions of contract of the principal contract. This is essential since the various NEC contracts include detailed powers and duties of the Adjudicator either in option W1 and W2 or, in the case of the ECSC and ECSS, in the core clauses. If, for some reason, there is any conflict between the two contracts, the AC conditions have precedence. The first clause repeats the basic obligation of the Adjudicator to act impartially; this is also stated in the principal contracts.

Expenses due to the Adjudicator are specified Because payments to the Adjudicator include expenses as well as his fee, it is important to state in the contract what the expenses are. Any expenses not listed are then deemed to be included in the Adjudicator's fee. One of the most important items is charges by others for help in the adjudication. The Adjudicator is entitled to seek help from others in order to reach his decision on a dispute. This help may be on matters of law or, perhaps, on some technical point on which he wishes to have an expert opinion. The conditions of contract require the Adjudicator to notify the parties prior to seeking such help and also to inform the parties of the advice received afterwards and invite their comments. There is obviously a cost involved and it is this cost that forms part of the Adjudicator's expenses.

A dispute may involve other parties

It is possible that a dispute in a contract under, say, the ECC involves work done by a Subcontractor under the ECS. Thus a dispute under the subcontract between the Contractor and Subcontractor may also be a dispute under the main contract between the Employer and Contractor, the subject matter of the disputes being the same in both contracts. It is clearly desirable that what is essentially the same dispute should not be referred to two different adjudicators who may produce different decisions. Accordingly, in the NEC contracts other than the ECSC and ECSS, there are 'joinder' clauses which allow for such a dispute to be decided between the three parties by the main contract Adjudicator. The AC makes provision for this procedure and states that reference to the parties are to be interpreted as including all three parties.

The AC describes the conduct of the adjudication in general terms only

Since the detailed adjudication procedures are described in each main contract between the parties, the procedures in the AC are stated in general terms only. There is, however, a requirement that the parties must co-operate with the Adjudicator and comply with his requests and directions. The Adjudicator is also required to keep documents related to the dispute for the period (the period of retention) stated in the contract data.

Payment of the Adjudicator's fee is on a time basis and the AC allows for an advance payment

The amount of any advanced payment is stated in the contract data. The Adjudicator's hourly rate is also stated in the contract data. Payment of the Adjudicator's fee is based on the time he spends on the adjudication, including time spent travelling. Payment becomes due within three weeks of the Adjudicator's invoices, which are submitted to both or possibly three parties. Payments are shared equally between the parties unless otherwise agreed. The parties are jointly and severally liable for the payment of fee and expenses. Thus, if one party fails to pay, the Adjudicator may recover the amount due from the other party or parties who then must seek to recover monies from the defaulting party.

The Adjudicator's appointment may be terminated by the parties or by the Adjudicator himself

The parties are entitled to terminate the Adjudicator's appointment for any reason. On the other hand, the Adjudicator is entitled to terminate for any of four stated reasons. The first of these is when a conflict of interest becomes apparent. In this connection, the Adjudicator has a duty to notify the parties as soon as he becomes aware of any conflict of interest. In any event, a long stop date for the termination of the Adjudicator's appointment is stated in the contract data. Under the various NEC contracts, disputes may arise at any time up to the end of the limitation period. Since it is unlikely that an adjudicator would wish to be bound for such a long period, provision has been made for earlier termination. If a dispute that the parties wish to refer to adjudication occurs after this date they will need to appoint a new adjudicator.

When to use the AC

Since its use is mandatory, the Adjudicator's Contract should be used for the appointment of an Adjudicator under any of the NEC contracts.

How to use the AC

Of all the NEC contracts, the AC is the easiest to prepare. Once the hourly fee rate has been agreed, the preparation of the contract documents is straightforward.

Exercises

Note: some additional knowledge will be required to answer the following.

(1) Discuss the merits of adjudication as a method of dealing with construction disputes. Compare and contrast adjudication with litigation in the courts, arbitration, conciliation, mediation and expert determination.

(2) Having failed to agree with the Contractor the assessment of a compensation event, the Project Manager decides its value at £26 000. The contract is in the UK and is under the ECC, using option A. It qualifies as a construction contract, as defined in the Housing Grants, Construction and Regeneration Act 1996. The Contractor believes that the Project Manager's assessment is much too low and is not prepared to accept it. Describe the procedure he should follow in order to have the dispute resolved.

Index

Page numbers in bold type refer to definitions. Abbreviations are those used in the text and are also referenced in the index.

AC see Adjudicators' Contracts
acceptance procedures, TSC 70
Activity Schedules **23, 56**
   payments for 11
   priced contracts with
      ECC 11
      PSC 49
   priced subcontracts with, ECS 30
   target contracts with, ECC 12–13
additional conditions see z clauses
Adjudicators
   ECC 22
   ECSC 38, 41
   introduction of 13
   PSC 51, 55
   rejection of decisions by 14
   TSC 69
Adjudicators' Contracts (AC)
   conditions of 76
   expenses and fees 76, 77
      advance payments 76
   forms of agreement 75
   joinder clauses 76
   origins 75
   and Subcontractor disputes 76
   termination of appointment 76–77
   use of 77
advanced payments
   to Contractors, ECC 17
   to subcontractors, ESC 32
Affected Properties **69**
amendments see z clauses

Banwell Report (1964) 4
Barnes, Dr Martin 1
Bills of Quantities **12, 23**
   priced contracts with 12
   priced subcontracts with 30
   target contracts with 13
   target subcontracts with 30
bonuses, early completion, ECC 15

cash flow, by lump sum payments 11
change, managing, ECSC 41
clause function statements 1–2
collateral warranty agreements, PSC 52
communications
   effective
      ECC 23
      ECSC 39–40
      PSC 57
      TSC 70
compensation

assessing
   ECC 24
   ECSC 37, 41
compensation events **22–23**
   Contractors 71
   ECC 24–25
   ECSC 40–41
   PSC 58–59
   TSC 71
Completion **22, 39, 55**
completion dates **55**
consequential losses, defects, ECC 18
'Constructing the Team' see Latham Report
construction management, contract selection 6
Consultants
   see also Professional Services Contract
   appointment of 4
Contractors
   advanced payments by Employers, ECC 17
   agents, ECSC 38
   compensation events 71
   defects, liability for 17, 70–71
   as designers, ECC 19–20
   Employers, effective communications 57
   insurance 71
   limitations of liability, ECC 17, 18
   non-performance by, ECC 17
   offers by 37
   roles and responsibilities
      ECC 20
      TSC 68
   subcontracting, permission for 68
   unfair risk allocation 10–11
Contractors' plans, TSC 70
Contractors' shares, calculation 12
contracts, creation of 36–37
Contracts (Rights of Third Parties) Act 1999
   ECC 19
   TSC 67
Cost Components
   Schedule of **23**
   Shorter **23**
cost reimbursable contracts
   ECC 12–13
   ECS 30
   TSC 64
Costs
   Defined **12, 23, 39**
   Disallowed **23**
currencies
   exchange rates 15
   multiple
      ECC 15

currencies (*continued*)
  ESC 31
  TSC 65

damages
  caps 16
  delays
    ECC 15–16
    ECS 32
  liquidated 15–16
  low performance
    ECC 18
    ECS 32
    TSC 66
  unliquidated 16
defects **22, 55–56**
  consequential losses, ECC 18
  Contractors' liabilities, ECC 17
  dates, ECSS 45
  management of
    ECC 24
    ECSC 40
    TSC 70–71
  notification, ECC 21, **22**
  retention, ECC 17–18
Defects Certificates 21, **22**
Defined Costs **12, 23, 39, 69**
delay damages
  ECC 15–16
    caps 16
  ECS 32
design work
  by Contractors, limitations of liability 17
  ECSC 38–39
  PSC 17
designers 19
  Contractors as, ECC 19–20
Disallowed Costs **23, 69**
dispute resolution
  Adjudicators 13–14
  ECC 13–14, 19
  ECSC 41
  ESC 30–31
  PSC 51, 55
  tribunals 14
  TSC 64

early completion bonuses
  ECC 15
  ESC 31
early take over, Employers, ECC 24
early warning procedures
  PSC 57
  TSC 70
ECC see Engineering and Construction Contracts
ECSC see Engineering and Construction Short
    Contracts
ECSS see Engineering and Construction Short
    Subcontracts
emergency works, contracts for 13
Employers
  acceptance by 37

advanced payments to Contractors, ECC 17
breach of contract by 58
Contractors, effective communications 57
duties, ECSC 38
early take over, ECC 24
inflation risks, ECC 14–15
limitations of liability, ECC 18
low performance damages, ECC 18
PSC, termination 52
rights and responsibilities, TSC 68
roles and duties
  ECC 20
  PSC 54–55
end of liability dates 66
Engineering and Construction Contracts (ECC)
  Adjudicators, appointment of 22
  advanced payments to Contractors 17
  changes in the law 15
  communications, format 23
  compensation
    assessing 24
    events 24–25
  defects
    Contractors' liability 17
    management of 24
    retention for 17–18
  delay damages 15–16
  development of 10
  dispute resolution 13–14, 19
  early completion bonuses 15
  and ECS 28
  Key Performance Indicators 18–19
  limitations of liability 17, 18
  low performance damages 18
  multiple currency payments 15
  options
    cost reimbursable contracts 13
    management contracts 13
    priced contracts with activity schedule 11
    priced contracts with bill of quantities 12
    target contracts with activity schedule 12–13
    target contracts with bill of quantities 13
  parent company guarantees 15
  partnering 16, 20
  performance bonds 17
  price adjustments, inflation 14–15
  programmes 24
  Project Managers, appointment 20–21
  risk reduction meetings 24
  roles within 19–22
  sectional completion 15
  structure 11
  Subcontractors, appointment 20, 21
  Supervisors, appointment 21
  tenders 25
  third party rights 19
  uses of 25–26
  z clauses 19
Engineering and Construction Short Contracts (ECSC)
  Adjudicators 38, 41
  communications, effective 39–40
  compensation assessments 37, 41

compensation events 40–41
contract creation 36–37
Contractors
    agents 38
    roles 39
defects, management of 40
design under 38–39
development of 36
dispute resolution 41
Employers, roles 39
payment procedures 40
Price Lists 37
problems, early warnings 40
Site Information 38
subcontracting 40
tender documents 42
uses of 41–42
Works Information 37–38
Engineering and Construction Short Subcontracts
    (ECSS)
Compensation Events 46
    weather 46
defect dates 45
development 44
and ECSC 44–45, 46
information required under 45
Price List pricing 45
and PSC 44–45
and TSC 44
uses of 46
Works Information 45
Engineering and Construction Subcontracts (ESC)
advanced payments to subcontractors 32
construction of 29–30
damages, delays 32
dispute resolution 30–31
early completion bonuses 31
and ECC 28
key dates 28
Key Performance Indicators 32
limitations of liability 32
low performance damages 32
multiple currency payments 31
options
    cost reimbursable subcontract 30
    priced subcontracts with Activity Schedule 30
    priced subcontracts with Bill of Quantities 30
    target subcontracts with Activity Schedules 30
    target subcontracts with Bills of Quantities 30
parent company guarantees 31
partnering 32
performance bonds 32
price adjustments for inflation 31
retention 32
sectional completion 31
subsubcontractors 33
suspension of performance 33
third party rights 33
time-barring sanctions 34
uses of 34
Works Information 28
z clauses 33

Equipment **22**, **39**, **69**
ESC see Engineering and Construction Subcontracts
exchange rates, currencies 15
expenses and fees, AC 76

Fees 63
freedom to contract 7

Gantt charts 38
'Guide to Project Team Partnering' 16

Housing Grants, Construction and Regeneration Act
    1996
    AC 75
    construction contracts **31**
    dispute resolutions
        ECC 13–14, 19, 21–22
        ECSC 41
        ESC 30–31
        PSC 51, 53
        TSC 64, 67

incentives, Key Performance Indicators as 16
inflation
    price adjustments
        ECC 14–15
        ESC 31
        PSC 51
        TSC 64–65
innovation, advantages and pitfalls of 2
insurance, Contractors 71

joinder clauses, AC 76

key dates **56**
    ESC 28
    sectional completion, ECC 15
Key Performance Indicators (KPI)
    ECC 16, 18–19
    ESC 32
    TSC 66–67

Latham Report (1994) 4–5
    adjudication 13
    mutual trust and co-operation 67
Latham, Sir Michael 4
    see also Latham Report (1994)
law changes
    ECC 15
    TSC 65
liability, end of 66
limitations of liability
    ECC 17, 18
    ESC 32
    PSC 53
    TSC 66
liquidated damages 15–16
low performance damages
    ECC 18
    ESC 32
    TSC 66
lump sum contracts **11**

lump sum items 63

management contracts, ECC, subcontracted work 13
multiple currencies
  ECC 15
  ESC 31
  TSC 65
mutual trust and co-operation 64, 67, 69

New Engineering Contracts (NEC)
  dates of publication 5
  feedback on 4
  objectives
    clarity 3
    flexibility 3
    simplicity 3
  options
    main 7
    secondary 7
    z clauses 8
  partnering within 16
  selection of appropriate 5–6
    construction management 6
    subcontracts 6
  terminology 3–4

parent company guarantees
  ECC 15
  ESC 31
  TSC 65
partnering
  ECC 16, 20
  ESC 32
  PSC 52, 54
  TSC 65–66
Partnering Information 16
  PSC 54
Partners, Schedule of 16
payment procedures, ECSC 40
performance bonds
  ECC 17
  ESC 32
  TSC 66
Perry, Professor John 1
Plant and Materials 39, 69
Price Lists 39
  ECSC 37, 45
  lump sum items 63
  remeasurable items 63
Priced Contracts with Activity Schedules
  ECC 11
  PSC 49
Priced Contracts with Price Lists, TSC 63
Prices for Services Provided to Date 56, 58, 69
Prices, The 23, 39, 56, 69
Prices for Work Done to Date 23, 39
Professional Services Contracts (PSC)
  Adjudicators 51, 55
  collateral warranty agreements 52
  communications, effective 57
  compensation events 58–59
  delegation under 52

development 48
dispute resolution 51, 55
early warning procedures 57
and ECSS 44
limits of liability 53
options
  priced contract with activity schedule 49
  target contracts 49–50
  term contracts 50
  time-based contracts 50
partnering 52, 54
  information 54
payments under 58
price adjustments, inflation 51
programmes within 57–58
roles and duties under 54
structure 48–49
subcontracting under 54–55
termination by employers 52
third party rights 53
uses of 59
z clauses 53
programmes
  ECC 24
  ECSC 38
  within PSC 57–58
Project Managers
  appointment
    ECC 20–21
    PSC 59
  delegation by, ECC 21
  introduction of 4
  qualifications, ECC 20
  subcontractors, approval of 20, 29
PSC see Professional Services Contract

quantity estimates, by employers 12

remeasurable items 63
retention
  ECSC 40
  ESC 32
risk reduction meetings
  ECC 24
  TSC 70
risk registers 7, 56
risks
  allocation 6–7
    to contractors 10–11

Schedule of Cost Components 23
  Shorter 23
Schedule of Partners 16
schedule of rates contracts see Term Service
  Contracts
scope 56
sectional completion
  key dates
    ECC 15
    ESC 31
Service Information 69
  TSC 72

Service Managers
  TSC
    appointment 62, 67
    Contractor's plans 70
    role of 68
    subcontracting 68
Shorter Schedule of Cost Components **23**
Site Information **22**
  ECS 34
  ECSC 38
standard conditions of contract, pre-NEC 1
Subconsultants, PSC 54–55
subcontract forms, need for 4
Subcontractors **29**
  see also Engineering and Construction Subcontract
  appointment of
    ECC 21, 29
    ECSC 40
  disputes with Contractors 76
  management contracts, ECC 13
  obligations 33–34
  partnership by 21
subcontracts
  selection of appropriate 6
  TSC as 71
subsubcontractors, ESC 33
Supervisors
  appointment of
    ECC 21
    PSC 59
  introduction of 4
suspension of performance, ESC 33

target contracts
  ECC 12–13
  ESC 30
  with price lists, TSC 63–64
  PSC 49–50
target prices, establishment of 13
tender programmes, ECC 24
tenders
  ECC 25
  ECSC 42
  Price List compilation, ECSC 37
term contracts
  PSC 50
    pricing 50
Term Service Contracts (TSC)
  acceptance procedures 70
  Adjudicator's role 69
  changes in the law 65
  compensation events 71
  construction 62
  Contractor's plans 70
  defect management 70–71
  development 61

dispute resolution 64, 67
early warnings 70
effective communication 70
Key Performance Indicators 66–67
limits of liability 66
low performance damages 66
multiple currency payments 65
mutual trust and co-operation 64, 67, 69
options
  cost reimbursable contracts 64
  priced contracts with price lists 63
  target contracts with price lists 63–64
parent company guarantees 65
partnering 65–66
performance bonds 66
price adjustments for inflation 64–65
roles and duties within 67–69
Service Information 72
Service Managers 62, 67
as subcontract 71
subcontracting within 68–69
time period 44
uses of 61, 71–72
z clauses 67
terminology, policies on 3–4
third party rights
  ECC 19
  ESC 33
  PSC 53
time charges **56**
time provisions, incorporation 3
time-barring sanctions, ESC 34
time-based contracts, PSC 50
transfer of rights, PSC 52
TSC see Term Service Contract

unliquidated damages 16

value engineering clauses 12

weather risks, ECSC 40–41, 46
working areas **22**
Works Information **22**
  ECSC 37–38
  ECSS 45
  ESC 28
  performance bonds, ECC 17

z clauses
  ECC 19
  ECS 33
  effects of 2
  flexibility using 8
  minimising use of 3
  PSC 53
  TSC 67